Deepwater Petroleum Exploration & Production

Deepwater Petroleum Exploration & Production

Editor

Gaurinath Mahatha

Deepwater Petroleum Exploration & Production

Edited by **Gaurinath Mahatha**

Printed in 2017

ISBN: 978-1-68117-352-8

Library of Congress Control Number: 2015939265

© 2016 by
SCITUS Academics LLC,
616, Corporate Way, Suite 2, 4766,
Valley Cottage, NY 10989

www.scitusacademics.com

This book contains information obtained from highly regarded resources. Copyright for individual articles remains with the authors as indicated. All chapters are distributed under the terms of the Creative Commons Attribution License, which permits unrestricted use, distribution, and reproduction in any medium, provided the original author and source are credited.

Notice

Reasonable efforts have been made to publish reliable data and views articulated in the chapters are those of the individual contributors, and not necessarily those of the editors or publishers. Editors or publishers are not responsible for the accuracy of the information in the published chapters or consequences of their use. The publisher believes no responsibility for any damage or grievance to the persons or property arising out of the use of any materials, instructions, methods or thoughts in the book. The editors and the publisher have attempted to trace the copyright holders of all material reproduced in this publication and apologize to copyright holders if permission has not been obtained. If any copyright holder has not been acknowledged, please write to us so we may rectify.

Contents

Preface vii

Chapter 1   Fracturing Pressure of Shallow Sediment in Deep Water Drilling 1

Chuanliang Yan, Jingen Deng, Lianbo Hu, and Baohua Yu

Chapter 2   Early Diagenesis Records and Pore Water Composition of Methane-Seep Sediments from the Southeast Hainan Basin, South China Sea 23

Daidai Wu, Nengyou Wu, Ying Ye, Mei Zhang, Lihua Liu, Hongxiang Guan, and Xiaorong Cong

Chapter 3   Assessment of the Deepwater Horizon Oil Spill Impact on Gulf Coast Microbial Communities 47

Regina Lamendella, Steven Strutt, Sharon Borglin, Romy Chakraborty, Neslihan Tas, Olivia U. Mason, Jenni Hultman, Emmanuel Prestat, Terry C. Hazen, and Janet K. Jansson

Chapter 4   Multiscale Erosion Surfaces of the Organic-Rich Barnett Shale, Fort Worth Basin, USA 81

Mohamed O. Abouelresh

Chapter 5   Flexible Riser Monitoring using Hybrid Magnetic/ Optical Strain Gage Techniques through RLS Adaptive Filtering 125

Daniel Pipa, Sérgio Morikawa, Gustavo Pires, Claudio Camerini, and JoãoMárcio Santos

Chapter 6   Improve the Government Strategic Petroleum Reserves 159

Xiucheng Dong, Zhongbing Zhou, and Hui Li

Chapter 7   Studies on the Evaporation Regulation Mechanisms of Crude Oil and Petroleum Products 171

Merv F. Fingas

Chapter 8	**Numerical Study of Oil/Water Separation by Ceramic Membranes in the Presence of Turbulent Flow** 193	

Tássia Mota Vieira, Josedite Saraiva de Souza, Enivaldo Santos Barbosa, Acto de Lima Cunha, Severino Rodrigues de Farias Neto, and Antonio Gilson Barbosa de Lima

Chapter 9	**Modeling, Simulation and Optimization of Continuous Gas Lift Systems for Deepwater Offshore Petroleum Production** 219	

J.N.M. de Souza, J.L. de Medeiros, A.L.H. Costa, and G.C. Nunes

Citations 263

Index 267

Preface

Deepwater Petroleum Exploration & Production provides the best-selling original, explaining the unique challenges of oil and gas exploration and production in the world's deepwater provinces and Petroleum Exploration and Production is an awareness level course designed for professionals associated with the oil and gas industry. The course content is designed to expose the participants to the full life cycle of the oil and gas industry.

Editor

Chapter 1

Fracturing Pressure of Shallow Sediment in Deep Water Drilling

Chuanliang Yan, Jingen Deng, Lianbo Hu, and Baohua Yu

State Key Lab of Petroleum Resources and Prospecting, China University of Petroleum, Beijing 102249, China

ABSTRACT

The shallow sediment in deep water has weak strength and easily gets into plastic state under stress concentration induced by oil and gas drilling. During drilling, the formation around a wellbore can be divided into elastic zone and plastic zone. The unified strength theory was used after yielding. The radius of the plastic zone and the theoretical solution of the stress distribution in these two zones were derived in undrained condition. The calculation model of excess pore pressure induced by drilling was obtained with the introduction of Henkel's excess pore pressure theory. Combined with hydraulic fracturing theory, the fracturing mechanism of shallow sediment was analyzed and the theoretical formula of fracturing pressure was given. Furthermore, the influence of the parameters of unified strength theory on fracturing pressure was analyzed. The theoretical calculation results agreed with measured results approximately, which preliminary verifies the reliability of this theory.

INTRODUCTION

With substantial exploration, the oil and gas resources on land and shallow water field cannot meet the need of the development of industry. Recently, more than 40% of the newly proved oil and gas resources are found in deep water field, which is much more difficult to develop [1]. Exploring and developing oil and gas in deep water field are one of the important developing trends of oil and gas industry.

Maintaining wellbore stability is an important issue in oil and gas industry. In the process of drilling, the economic losses caused by wellbore instability reach more than one billion dollar every year [2], and the lost time is accounting for over 40% of all drilling related nonproductive time [3]. For deep water drilling, the overburden pressure and fracturing pressure are lower than that on land, so the risk of wellbore instability is much higher [4]. The conventional calculation models of fracturing pressure are based on the elastic mechanics [5–9]. However, in some cases, the formation experiences plastic before the fracture initiates. Aadnøy and Belayneh [10] established an elastic-plastic model for the calculation of fracturing pressure with the consideration of the plastic zone around a wellbore. According to his theory, the fracture initiates at the interface of the elastic zone and the plastic zone, which leads to the calculation result much higher than that of the elastic model and disagrees with the fact that the fracturing pressure of shallow sediment in deep water is low. For deep water drilling, researchers mostly analyze the fracturing pressure based on the elastic model or empirical model based on the field testing data [11–15]; this may be suitable for the deep and hard layer but cannot be applied to the shallow sediment in deep water, which is unconsolidated, subjected to the consolidation theory of saturated soil, and experiences plastic before fracturing [1, 16, 17]. The traditional models cannot reveal its fracturing mechanism. Although some researchers have done research on the fracturing pressure of shallow sediment in deep water specially [1, 16, 18–20], these models are empirical or semiempirical models based on the field testing data and they are not based on the stress state around the wellbore and thus cannot explain the fracturing mechanism of shallow sediment in deep water. The traditional wellbore stability analysis is set up on the basis of rock mechanics engineering. Excess pore pressure theory [21] in soil mechanics will be adopted in this paper. Formations around the wellbore are divided into elastic zone and plastic zone. The formation in plastic zone obeys the unified strength theory. Considering the excess pore pressure due to soil squeezing effect induced by wellbore column pressure, combined with hydraulic fracturing theory to research the fracturing mechanism of shallow sediment in deep water. Assuming that tensile stress will appear when the pore pressure is greater than the combination of the insitu stress and the increment of stress due to drilling. The fracturing pressure is defined as the wellbore column pressure when the tensile stress exceeds the tensile strength of the formation when the fracture initiates.

MECHANICAL MODEL AND BASIC HYPOTHESIS

The shallow sediment in deep water can be treated as homogeneous and isotropic ideal elastoplastic material [16, 22–24]. The sediment is in elastic state before drilling and obeys the unified strength theory after yielding. Shallow sediment in deep water experienced short deposition time and little tectonic movement, and the Poisson's ratio is large, so the difference of the horizontal stresses is very small. It is reasonable to assume that the horizontal in-situ stress of shallow sediment in deep water is uniform [16, 20].

The formation around the wellbore is subjected to the in-situ stress and the drilling fluid column pressure. The stress field around the wellbore redistributes and stress concentration occurs while drilling. When the stress state around the wellbore exceeds the elastic limitation, the near wellbore formation will turn into plastic state, while the formation beyond the plastic zone will still remain elastic. The mechanical model of a wellbore is shown in Figure 1.

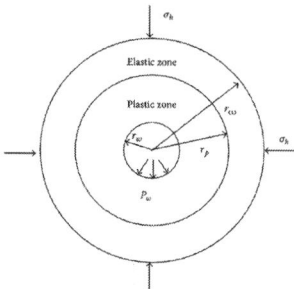

Figure 1: The mechanical model of a wellbore.

As shown in Figure 1, σ_h is the far-field horizontal in-situ stress; P_{p0} is the original pore pressure; P_p is the pore pressure after drilling; P_w is the wellbore pressure; r_w is the wellbore radius; r_p is the radius of the elastic-plastic zone interface, so the plastic zone is within the radius of r_p; the formation which is outside of the radius r_p is still in elastic state. The unified strength theory was proposed by Yu and He in 1991 [25]. This theory has clear physical concepts and simple mathematical formula and agrees with experimental results greatly, and it performed well in various engineering all over the world [26–29]. This theory is based on the physical model of twin shear stress element. This theory not only takes into consideration the different influence of shear stress and the normal stress on the yielding and failure of materials but also the influence of intermediate principle stress in different stress state. So this theory can be applied to a large range of materials from metal to rock material. When φ and C stand for internal friction angel and cohesive, respectively, the unified strength theory can be written

$$\sigma_1 = \frac{2(1+b)(1+\sin\varphi) + mb(\sin\varphi - 1)}{[2(1+b) - mb](1 - \sin\varphi)} \sigma_3$$

$$+ \frac{4(1+b)C\cos\varphi}{[2(1+b) - mb](1 - \sin\varphi)}, \quad (1)$$

As where b ($0 \leq b \leq 1$) reveals the influence of intermediate principle stress on the yielding and failure of materials. It is the parameter that reveals the influence of the intermediate stress and different strength criterions: when $b=0$, (1) is Mohr-Coulomb strength criterion; when $b=1$, the formula is twin shear strength criterion. m is the coefficient of intermediate stress whose physical meaning can be seen in Yu Maohong's research [30]. For plane strain state, $m \leq 1$; in elastic state, $m < 1$ in plastic state, m approaches 1; in this paper, m is assumed to be 1 when the shallow sediment is in plastic state. C, φ, are the sediment cohesion and internal friction angle in undrained condition, respectively; this paper uses the total stress shear strength parameters [31, 32].

For wellbore wall fracturing problem while drilling, it is always to be true that [9, 13] $\sigma_r = \sigma_1$ and $\sigma_\theta = \sigma_3$ the intermediate stress is presented as follows [13, 33–36]:

$$\sigma_z = \sigma_2 = \frac{m(\sigma_1 + \sigma_3)}{2}. \tag{2}$$

According to the unified strength theory, the yielding function of shallow sediment in deep water before fracturing can be expressed as follows:

$$\sigma_r = M\sigma_\theta + \sigma_0, \tag{3}$$

in which

$$M = \frac{2(1+b)(1+\sin\varphi) + mb(\sin\varphi - 1)}{[2(1+b) - mb](1 - \sin\varphi)},$$

$$\sigma_0 = \frac{4(1+b)C\cos\varphi}{[2(1+b) - mb](1 - \sin\varphi)}. \tag{4}$$

In the drilling process, a tight layer of mud cake will generate on the wellbore wall which can prevent the seepage of the fluid between the wellbore and the formation. So it is reasonable to assume that the shallow sediment is in untrained condition while drilling. The stress state around the wellbore will change due to drilling; under the undrained condition, part of the stress will be taken by the pore fluid, so the pore pressure will change after drilling. The change of the pore pressure caused by drilling is defined as excess pore pressure, marked with ΔP and expressed in the following equation:

$$\Delta P = P_p - P_{p0}. \tag{5}$$

FRACTURING PRESSURE OF SHALLOW SEDIMENT IN DEEP WATER

Elastic-Plastic Analysis around the Wellbore

According to the stress balance formula, the stress relationship at any point in the shallow sediment while drilling can be expressed as follows:

$$\frac{d\sigma_r}{dr} + \frac{\sigma_r - \sigma_\theta}{r} = 0, \tag{6}$$

where σ_r and σ_θ are the radial stress, and tangential stress respectively. In the elastic zone, according to the axial symmetry problems solution and boundary conditions: $\sigma_r|_{r=rp} = \sigma_{rp}$, $\lim r \to \infty \sigma_r = \sigma_h$, the near wellbore stress in the elastic zone is given by the following [37]:

$$\sigma_r = \sigma_h - (\sigma_h - \sigma_{rp})\frac{r_p^2}{r^2},$$

$$\sigma_z = \sigma_v,$$

$$\sigma_\theta = \sigma_h + (\sigma_h - \sigma_{rp})\frac{r_p^2}{r^2}, \tag{7}$$

where σ_{rp} is the radial stress at the interface of the plastic zone and the elastic zone. Combining (3) and (6) and using the boundary condition $\sigma_r|_{r=rw} = P_w$, the stress field in the plastic zone can be deduced and written as follows:

$$\sigma_r = \left(P_w + \frac{\sigma_0}{M-1}\right)\left(\frac{r_w}{r}\right)^{(M-1)/M} - \frac{\sigma_0}{M-1},$$

$$\sigma_z = \frac{M+1}{2M}\left(P_w + \frac{\sigma_0}{M-1}\right)\left(\frac{r_w}{r}\right)^{(M-1)/M} - \frac{\sigma_0}{M-1},$$

$$\sigma_\theta = \frac{1}{M}\left(P_w + \frac{\sigma_0}{M-1}\right)\left(\frac{r_w}{r}\right)^{(M-1)/M} - \frac{\sigma_0}{M-1}. \tag{8}$$

According to the continuity of the stress, the stress at the elastic-plastic interface should agree with both (7) and (8). So the radius of the plastic zone can be written as follows:

$$r_p = r_w \left[\frac{(M+1)\left[(M-1)P_w + \sigma_0\right]}{2M\left[(M-1)\sigma_h + \sigma_0\right]}\right]^{M/(M-1)} \tag{9}$$

The stresses at the elastic-plastic interface are

$$\sigma_{rp} = \frac{2M\sigma_h + \sigma_0}{M+1},$$

$$\sigma_{\theta p} = \frac{2\sigma_h - \sigma_0}{M+1}. \tag{10}$$

Inserting (9) and (10) into (7), the final stress solutions in the elastic zone can be written as

$$\sigma_r = \sigma_h - \frac{(1-M)\sigma_h - \sigma_0}{M+1}\frac{r_w^2}{r^2}$$

$$\times \left[\frac{(M+1)\left[(M-1)P_w + \sigma_0\right]}{2M\left[(M-1)\sigma_h + \sigma_0\right]}\right]^{2M/(M-1)},$$

$$\sigma_\theta = \sigma_h + \frac{(1-M)\sigma_h - \sigma_0}{M+1}\frac{r_w^2}{r^2}$$

$$\times \left[\frac{(M+1)\left[(M-1)P_w + \sigma_0\right]}{2M\left[(M-1)\sigma_h + \sigma_0\right]}\right]^{2M/(M-1)}. \tag{11}$$

Solution of Excess Pore Pressure

Saturated soil is composed of solid particle skeleton and pores full of water. When imposed upon by an external force, the force will be balanced by both pore pressure and effective stress; the pore pressure increment induced by external load is called "excess pore pressure" [21]. Drilling will result in stress concentration around the wellbore; under the undrained condition, it will cause the change of pore pressure. At present, the solution of excess pore pressure is based on the research of Skempton and Henkel [38]. Skempton [39] established the calculation formula of excess pore pressure based on the experimental research of conventional triaxial tests of the soil:

$$\Delta P = B\left[\Delta\sigma_3 + A\left(\Delta\sigma_1 - \Delta\sigma_3\right)\right], \tag{12}$$

where B is the Skempton's pore pressure parameter under the acting of both isotropic stress and deviatoric stress, and it is related to the soil saturation, for saturated soil, $B = 1.0$; A is the Skempton's pore pressure parameter under the acting of deviatoric stress, which can be determined by experiment or experience (as shown in Table 1). σ_1 and $\Delta\sigma_3$ stand for the maximum and the minimum principal stress increments, respectively

Table 1: The experience value of A [40]

Soil type (saturated)	A
Loose fine sand	2~3
Sensitive clay	0.75~1.5
Normally consolidated clay	0.5~1.0
Mildly consolidated clay	0~0.5
Seriously consolidated clay	−0.5~0

Henkel [41] reasoned that under complex stress condition, the calculation of excess pore pressure should take the influence of the intermediate principle stress into consideration. The excess pore pressure under triaxial stress condition consists of two parts: one part caused by mean normal stress and the other part caused by mean shear stress. He put forward the following calculation formulas:

$$\Delta P = \beta \Delta \sigma_{oct} + \alpha \Delta \tau_{oct}, \qquad (13)$$

in which

$$\Delta \sigma_{oct} = \frac{1}{3}(\Delta \sigma_1 + \Delta \sigma_2 + \Delta \sigma_3),$$

$$\Delta \tau_{oct} = \frac{1}{3}\sqrt{(\Delta \sigma_1 - \Delta \sigma_2)^2 + (\Delta \sigma_2 - \Delta \sigma_3)^2 + (\Delta \sigma_3 - \Delta \sigma_1)^2}, \qquad (14)$$

where α and β are the Henkel's pore pressure coefficients. For saturated soil, $\beta=1$. In conventional triaxial tests, it has the relationship $\Delta \sigma_2 = \Delta \sigma_3$. Inserting it to the Henkel's excess pore pressure formula, the following formula is available:

$$\Delta P = \Delta \sigma_3 + \frac{\left(1 + \alpha \sqrt{2}\right)(\Delta \sigma_1 - \Delta \sigma_3)}{3}. \qquad (15)$$

Compared with Skempton's formula, Henkel's excess pore pressure formula can be rewritten as follows:

$$\Delta P = \beta \Delta \sigma_{oct} + \frac{(3A - 1)\Delta \tau_{oct}}{\sqrt{2}}$$

$$\approx \beta \Delta \sigma_{oct} + (2.12A - 0.71)\Delta \tau_{oct}. \qquad (16)$$

According to (8), the stress increment in the plastic zone induced by drilling is expressed as follows:

$$\Delta \sigma_r = \left(P_w + \frac{\sigma_0}{M-1}\right)\left(\frac{r_w}{r}\right)^{(M-1)/M} - \frac{\sigma_0}{M-1} - \sigma_h,$$

$$\Delta \sigma_z = \frac{M+1}{2M}\left(P_w + \frac{\sigma_0}{M-1}\right)\left(\frac{r_w}{r}\right)^{(M-1)/M} - \frac{\sigma_0}{M-1} - \sigma_h,$$

$$\Delta \sigma_\theta = \frac{1}{M}\left(P_w + \frac{\sigma_0}{M-1}\right)\left(\frac{r_w}{r}\right)^{(M-1)/M} - \frac{\sigma_0}{M-1} - \sigma_h. \qquad (17)$$

Calculating the mean normal stress increment $\Delta \tau_{oct}$ and the mean shear stress increment $\Delta \tau_{oct}$ based on (17) and then inserting them into (16), the excess pore pressure in the plastic zone will be obtained by Henkel's excess pore pressure formula:

$$\Delta P = \frac{(1.73A+0.42)(M-1)+2}{2M}\left(P_w + \frac{\sigma_0}{M-1}\right)\left(\frac{r_w}{r}\right)^{(M-1)/M}$$

$$- \frac{\sigma_0}{M-1} - \sigma_h. \tag{18}$$

The principle stress increment in the elastic zone can be deduced based on the stress distribution equation (11). One has

$$\Delta\sigma_r = -\frac{(1-M)\sigma_h - \sigma_0}{M+1}\frac{r_w^2}{r^2}$$

$$\times \left[\frac{(M+1)\left[(M-1)P_w + \sigma_0\right]}{2M\left[(M-1)\sigma_h + \sigma_0\right]}\right]^{2M/(M-1)},$$

$$\Delta\sigma_z = 0,$$

$$\Delta\sigma_\theta = \frac{(1-M)\sigma_h - \sigma_0}{M+1}\frac{r_w^2}{r^2}$$

$$\times \left[\frac{(M+1)\left[(M-1)P_w + \sigma_0\right]}{2M\left[(M-1)\sigma_h + \sigma_0\right]}\right]^{2M/(M-1)}. \tag{19}$$

The excess pore pressure in the elastic zone calculated by Henkel's excess pore pressure formula can be written as

$$\Delta P = \frac{(0.58 - 1.73A)\left[(1-M)\sigma_h - \sigma_0\right]}{M+1}\frac{r_w^2}{r^2}$$

$$\times \left[\frac{(M+1)\left[(M-1)P_w + \sigma_0\right]}{2M\left[(M-1)\sigma_h + \sigma_0\right]}\right]^{2M/(M-1)} \tag{20}$$

Effective Stress around the Wellbore

According to Terzaghi's effective stress principle, the effective stress in the plastic zone can be written as follows:

$$\sigma'_r = \sigma_r - \Delta P - P_{p0}$$
$$= \frac{(1.58 - 1.73A)(M-1)}{2M}\left(P_w + \frac{\sigma_0}{M-1}\right)\left(\frac{r_w}{r}\right)^{(M-1)/M}$$
$$+ \sigma_h - P_{p0},$$

$$\sigma'_\theta = \sigma_\theta - \Delta P - P_{p0}$$
$$= \frac{-(1.73A + 0.42)(M-1)}{2M}\left(P_w + \frac{\sigma_0}{M-1}\right)\left(\frac{r_w}{r}\right)^{(M-1)/M}$$
$$+ \sigma_h - P_{p0},$$

$$\sigma'_z = \sigma_z - \Delta P - P_{p0}$$
$$= \frac{(0.58 - 1.73A)(M-1)}{2M}\left(P_w + \frac{\sigma_0}{M-1}\right)\left(\frac{r_w}{r}\right)^{(M-1)/M}$$
$$+ \sigma_h - P_{p0}. \tag{21}$$

The effective stress in the elastic zone can be written as

$$\sigma'_r = \sigma_h - \frac{(1.58 - 1.73A)\left[(1-M)\sigma_h - \sigma_0\right]}{M+1}\frac{r_w^2}{r^2}$$
$$\times \left[\frac{(M+1)\left[(M-1)P_w + \sigma_0\right]}{2M\left[(M-1)\sigma_h + \sigma_0\right]}\right]^{2M/(M-1)} - P_{p0},$$

$$\sigma'_\theta = \sigma_h + \frac{(0.42 + 1.73A)\left[(1-M)\sigma_h - \sigma_0\right]}{M+1}\frac{r_w^2}{r^2}$$
$$\times \left[\frac{(M+1)\left[(M-1)P_w + \sigma_0\right]}{2M\left[(M-1)\sigma_h + \sigma_0\right]}\right]^{2M/(M-1)} - P_{p0},$$

$$\sigma'_z = \sigma_v - \frac{(0.58 - 1.73A)\left[(1-M)\sigma_h - \sigma_0\right]}{M+1}\frac{r_w^2}{r^2}$$
$$\times \left[\frac{(M+1)\left[(M-1)P_w + \sigma_0\right]}{2M\left[(M-1)\sigma_h + \sigma_0\right]}\right]^{2M/(M-1)} - P_{p0}. \tag{22}$$

Fracturing Pressure

When the pore pressure in shallow sediment unit is equal to the external pressure of that unit, the soil is under critical condition. If the pore pressure exceeds the critical value, the effective stress in soil will become a tensile stress. The fracturing pressure is defined as the wellbore pressure when the tensile stress exceeds the tensile strength of the formation.

The boundary condition on the wellbore wall should be satisfied when fracture initiates [37, 42]:

$$\sigma'_\theta \big|_{r=r_w} = -S_t, \tag{23}$$

where S_t is the tensile strength of the formation.

According to (21), the tangential effective stress on the wellbore wall can be written as follows:

$$\sigma'_\theta \big|_{r=r_w} = \frac{-(1.73A + 0.42)(M-1)}{2M}\left(P_w + \frac{\sigma_0}{M-1}\right) + \sigma_h - P_{p0}. \tag{24}$$

Inserting (24) into (23), the calculation formula of fracturing pressure for shallow sediment in deep water can be got:

$$P_f = \frac{2M(\sigma_h - P_{p0} + St)}{(1.73A + 0.42)(M-1)} - \frac{\sigma_0}{M-1}, \tag{25}$$

where P_f is the fracturing pressure.

ANALYSIS

To research the variation rule of the fracturing pressure, we took a set of formation parameters to analyze. The parameters are as follows: σ_h = 26 MP$_a$, P_{p0} = 21 MP$_a$, C = 0.2 MPa, φ = 4.5°, S_t = 0.02 MP$_a$, E = 300 MP$_a$, V = 0.35, and A = 0.7. The relationship of the fracturing pressure and the intermediate stress coefficient "b" is shown in Figure 2.

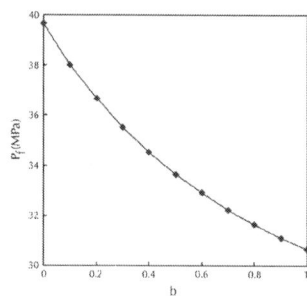

Figure 2: Variation of the fracturing pressure with the intermediate stress coefficient "b."

As shown in Figure 2, it is obvious that the fracturing pressure decreases nonlinearly with the increasing of coefficient "b." When b=1, the fracturing pressure decreases by approximate 23% compared with the value when .b=0

With the parameters listed above, the change rules of the tangential stress and the excess pore pressure on the wellbore wall were analyzed and shown in Figure 3 to Figure 5.

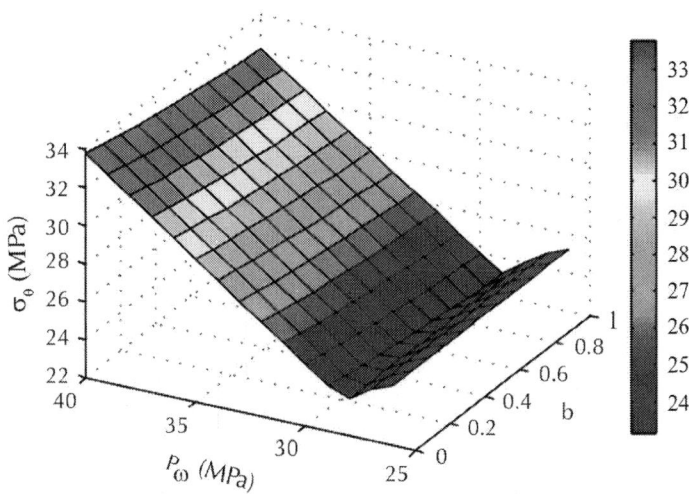

Figure 3: The relationship of tangential total stress on the wellbore wall with "b" and p_w.

Figure 3 shows the change rule of the tangential total stress on the wellbore wall. When the wellbore pressure is less than 29 MPa, the wellbore wall is in elastic state. The tangential total stress decreases with the increasing of the wellbore pressure increasing in elastic state, but in plastic state it increases with the increasing of the wellbore pressure and decreases with the increasing of the coefficient "b."

Figure 4 shows the change of the excess pore pressure with the wellbore pressure and the coefficient "b." It reveals that higher drilling fluid density or smaller coefficient "b" means a higher excess pore pressure.

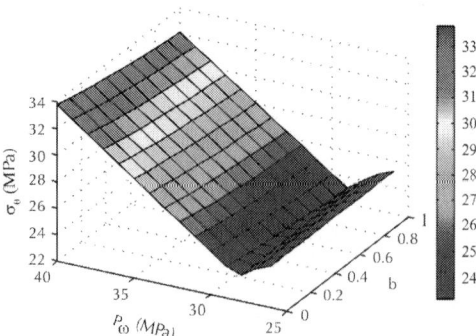

Figure 4: The relationship of excess pore pressure on the wellbore wall with "b" and p_w.

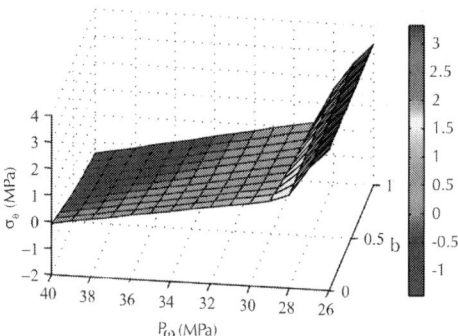

Figure 5: The relationship of tangential effective stress on the wellbore wall with "b" and p_w.

Figure 5 shows the variation of tangential effective stress. It reveals that the change rule of tangential effective stress is different from that of the total stress. The tangential effective stress always decreases with the increasing of the wellbore pressure and decreases with the coefficient "b" increasing. It is the influence of excess pore pressure that causes the great difference between the change rules of total stress and effective stress. The increasing rate of the excess pore pressure is higher than that of the total stress, so even though the tangential stress in plastic state increases with a wellbore pressure increasing, the effective stress decreases gradually.

CASE STUDY

To verify the theory in this paper, the field measured results of a deep water oil field at West Africa are taken as a case study. The calculation parameters are listed in Table 2 and the calculation results and measured results are shown in Table 3. The actual measured fracturing pressure was obtained from the leak off tests (LOT).

Table 2: The calculation parameters of fracturing pressure

Well number	Water depth/m	Formation depth/m	σ_h / MPa	/MP$_a$	$p_p°$/MP$_a$	φ /°	/MP$_a$	A
A1	1314	1868	23.1	19.2	0.1	4.4	0.02	0.7
A2	1360	2010	25.3	21.0	0.1	4.7	0.02	0.7
A3	1432	1987	24.4	20.5	0.1	4.3	0.02	0.7

Table 3: The comparison of theoretical results and LOT results

b	Theoretical results (MPa)			LOT results (MPa)		
	A1	A2	A3	A1	A2	A3
0	32.3	36.1	33.9			
0.1	31.0	34.7	32.5			
0.2	29.9	33.5	31.4			
0.3	29.0	32.4	30.4			
0.4	28.2	31.6	29.6			
0.5	27.5	30.8	28.9	26.1	28.8	27.6
0.6	26.9	30.2	28.2			
0.7	26.4	29.6	27.7			
0.8	25.9	29.1	27.2			
0.9	25.5	28.6	26.7			
1	25.1	28.2	26.3			

As shown in Table 3, when $b=1$, the error of the theoretical value for well A1 compared with the measured value is 3.8%, and for well A2 and well A3, the error is 2.1% and 4.7%, respectively. The fracturing pressure increases with the coefficient "b" decreasing. This table shows a fairly good agreement between the calculation results and the measured results, which verifies the feasibility of introducing the excess pore pressure theory into the fracturing mechanism analysis of shallow sediment in deep water, and the influence of intermediate stress should be taken into account when calculating the fracturing pressure of shallow sediment in deep water.

CONCLUSIONS

- The shallow sediment in deep water is unconsolidated and easily getting into plastic state, so there must be a plastic zone before the fracture initiates. When calculating the fracturing pressure, the stress distribution rule around the wellbore after turning into plastic state needs to be analyzed.
- During the drilling process, a high wellbore pressure will cause soil squeezing effect near the wellbore and result in an increase of the pore pressure, so the excess pore pressure generates. The fracturing of shallow sediment is mainly due to the generation of excess pore pressure which makes the effective stress on the wellbore wall decrease to tensile stress.
- The fracturing pressure of shallow sediment in deep water decreases nonlinearly with the increasing of the intermediate principle stress coefficient "." In elastic state, the tangential total stress decreases with the increasing of the wellbore pressure; in plastic state, the tangential total stress increases with the increasing of the wellbore pressure and decreases with the increasing of "." The excess pore pressure increases with the drilling fluid density increasing and decreases little with "" increasing. The tangential effective stress decreases with the drilling fluid density increasing and decreases with "" increasing.
- The agreement of the theoretical results and the measured data verifies the feasibility of the theory established in this paper. The introduction of excess pore pressure theory into wellbore stability analysis successfully explains the fracture mechanism of shallow sediment in deep water, which is an innovative progress in the theory exploration.

- The shallow sediment in deep water has not changed to rock, and it is under the control of saturated soil consolidation theory. In the analysis of geological mechanics problems in shallow sediments, the soil mechanics theory should be introduced into the analysis.

ACKNOWLEDGMENTS

This paper is supported by Science Fund for Creative Research Groups of the National Natural Science Foundation of China (Grant no. 51221003), National Natural Science Foundation Project of China (Grant no. 51174219), and National Oil and Gas Major Project of China (Grant no. 2011ZX05009-005 and Grant no. 2011ZX05026-001-01).

REFERENCES

1. B. Yu, C. Yan, J. Deng, S. Liu, Q. Tan, and K. Xiao, "Evaluation and application of wellbore stability in deep water," Drilling & Production Technology, vol. 33, no. 6, pp. 1–4, 2011. .2
2. M. E. Zeynali, "Mechanical and physico-chemical aspects of wellbore stability during drilling operations," Journal of Petroleum Science and Engineering, vol. 82-83, pp. 120–124, 2012. .3
3. J. Zhang, J. Lang, and W. Standifird, "Stress, porosity, and failure-dependent compressional and shear velocity ratio and its application to wellbore stability," Journal of Petroleum Science and Engineering, vol. 69, no. 3-4, pp. 193–202, 2009. .4
4. S. M. Willson, S. Edwards, P. D. Heppard et al., "Wellbore stability challenges in the deep water, Gulf of Mexico: case history examples from the pompano field," in SPE Annual Technical Conference and Exhibition, Proceedings-Mile High Meeting of the Minds, pp. 1833–1842, October 2003, SPE 84266. .5
5. E. B. Eaton, "Fracture gradient prediction and its application in oilfield operations," Journal of Petroleum Technology, vol. 21, no. 10, pp. 1353–1360, 1969. .6
6. R. A. Anderson, D. S. Ingram, and A. M. Zanier, "Determining

fracture pressure gradients from well logs," Journal of Petroleum Technology, vol. 25, pp. 1259–1268, 1973. .7

7. H. Rongzun, "A model for predicting formation fracture pressure," Journal of East China Petroleum Institute, vol. 4, pp. 335– 347, 1984. .8

8. K. Guo and P. Chang, "Study on prediction of fracturing pressure of shallow layer," Chinese Journal of Rock Mechanics and Engineering, vol. 23, no. 14, pp. 2484–2487, 2004. .9

9. J. Huang, D. V. Griffiths, and S.-W. Wong, "Initiation pressure, location and orientation of hydraulic fracture," International Journal of Rock Mechanics and Mining Sciences, vol. 49, pp. 59– 67, 2012. .10

10. B. S. Aadnøy and M. Belayneh, "Elasto-plastic fracturing model for wellbore stability using non-penetrating fluids," Journal of Petroleum Science and Engineering, vol. 45, no. 3-4, pp. 179–192, 2004. .11

11. B. S. Aadnoy, "Geomechanical analysis for deep-water drilling," in Proceedings of the IADC/SPE Drilling Conference, pp. 441–456, March 1998, SPE 39339. .12

12. R. Chhajlani, Z. Zheng, D. Mayfield, and B. MacArthur, "Utilization of Geomechanics for Medusa Field Development, Deepwater Gulf of Mexico," in Proceedings of the SPE Annual Technical Conference and Exhibition, pp. 3821–3832, October 2002, SPE 77779. .13

13. D. G. Ritter and B. Grollimund, "Wellbore stability in the deepwater Bijupira & Salema fields, Offshore Brazil—a probabilistic approach," in Offshore Technology Conference (OTC '03), Houston, Tex, USA, May 2003, OTC 15205. .14

14. S. M. Willson, S. T. Edwards, A. Crook et al., "Assuring stability in extended-reach wells—analyses, practices, and mitigations," in SPE/IADC Drilling Conference and Exhibition 2007, pp. 358–371, February 2007, SPE 105405. .15

15. J. Lang, S. Li, and J. Zhang, "Wellbore stability modeling and real-time surveillance for deepwater drilling to weak bedding planes and depleted reservoirs," in SPE/IADC Drilling Conference and Exhibition 2011, pp. 145–162, March 2011, SPE/IADC 139708-MS. .16

16. K. Wojtanowicz, A. T. Bourgoyne, D. Zhou, and K. Bender, "Strength and fracture gradients for shallow marine sediments," Final Report Submitted To the U.S., Department of Interior Mineral Management Service, 2000. .17
17. L. S. Rocha, P. Junqueira, and J. Roque, "Overcoming deep and ultra deepwater drilling challenges," in Offshore Technology Conference (OTC '03), Houston, Tex, USA, May 2003, OTC 15233. .18
18. E. Kaarstad and B. S. Aadnoy, "Fracture model for general offshore applications," in SPE Asia Pacific Oil and Gas Conference and Exhibition 2006: Thriving on Volatility, pp. 867–872, September 2006, SPE 101178. .19
19. E. Karstad and B. S. Aadnoy, "Improved prediction of shallow ° sediment fracturing for offshore applications," SPE Drilling & Completion, vol. 23, no. 2, pp. 88–92, 2008. 8 Mathematical Problems in Engineering .20
20. B. Yu, C. Yan, J. Deng et al., "Study and application of calculation model of safe drilling fluid density window: a case study of AKPO oilfield, West Africa," China Offshore Oil and Gas, vol. 24, no. 2, pp. 58–60, 2012. .21
21. Z. Hu, Soil Mechanics and Environment Soil Engineering, Tongji University Press, Shanghai, China, 1997. .22
22. D.-S. Jeng, B. R. Seymour, and J. Li, "A new approximation for pore pressure accumulation in marine sediment due to water waves," International Journal for Numerical and Analytical Methods in Geomechanics, vol. 31, no. 1, pp. 53–69, 2007. .23
23. B. Dugan, "Fluid flow in the Keathley Canyon 151 Mini-Basin, northern Gulf of Mexico," Marine and Petroleum Geology, vol. 25, no. 9, pp. 919–923, 2008. .24
24. T.-H. Kwon, K.-I. Song, and G.-C. Cho, "Destabilization of marine gas hydrate-bearing sediments induced by a hot wellbore: a numerical approach," Energy & Fuels, vol. 24, no. 10, pp. 5493–5507, 2010. .25
25. M. Yu and L. He, "A new model and theory on yield failure of materials under the complex stress state," in Mechanical Behaviour of Materials S26, M. Jono and T. Inoue, Eds., pp. 841–846, Prergamon Press, Oxford, UK, 1991. .26

26. V. A. Kolupaev, M. Moneke, and F. Becker, Mehraxiales Kriechen von Thermoolast2Formteilen, VDI Verlag, Darmstadt, Germany, 2005. .27
27. H. Y. Liu, S. Q. Kou, P.-A. Lindqvist, and C. A. Tang, "Numerical simulation of the rock fragmentation process induced by indenters," International Journal of Rock Mechanics and Mining Sciences, vol. 39, no. 4, pp. 491–505, 2002. .28
28. J. Sun and S. Wang, "Rock mechanics and rock engineering in China: developments and current state-of-the-art," International Journal of Rock Mechanics and Mining Sciences, vol. 37, no. 3, pp. 447–465, 2000. .29
29. M.-H. Yu, "Advances in strength theories for materials under complex stress state in the 20th century," Applied Mechanics Reviews, vol. 55, no. 3, pp. 169–218, 2002. .30
30. M.-H. Yu, Unified StrengthTheory and Its Applications, Springer, Berlin, Germany, 2004. .31
31. T. Ramamurthy, "Shear strength response of some geological materials in triaxial compression," International Journal of Rock Mechanics and Mining Sciences, vol. 38, no. 5, pp. 683–697, 2001. .32
32. X.-N. Gong, "Some problems concerning shear strength of soil in soft clay ground,"Chinese Journal of Geotechnical Engineering, vol. 33, no. 10, pp. 1596–1600, 2011. .33
33. C.-G. Zhang, Q.-H. Zhang, and J.-H. Zhao, "Unified solutions of well-bore stability considering strain softening and shear dilation," Journal of the China Coal Society, vol. 34, no. 5, pp. 634–639, 2009. .34
34. J. Li, M. Yu, and S. Wang, "Unified limit analysis of a wellbore under the effect of pore pressure and seepage," Journal of Mechanical Strength, vol. 23, no. 2, pp. 239–242, 2001. .35
35. J. Wang, J. Zhao, L. Wang, J. Wang, and S. Sun, "Stress analysis of wellbore rock based on unified strength theory," Journal of Architecture and Civil Engineering, vol. 26, no. 3, pp. 105–109, 2009. .36
36. Z. Nie, B. Xia, L. Zhou, and J. Deng, "Modeling of wellbore stability for gas drilling," Natural Gas Industry, vol. 31, no. 6, pp. 71–76, 2011. .37

37. E. Fjær, R. M. Holt, P. Horsrud, A. M. Raaen, and R. Risnes, Petroleum Related Rock Mechanics, Elsevier, 2nd edition, 2008. .38
38. X. Gong, Advanced Soil Mechanics, Zhejiang University Press, Zhejiang, China, 1996. .39
39. W. Skempton, "The pore-pressure coefficients A and B," Geotechnique, vol. 4, no. 4, pp. 143–147, 1954. .40
40. J. Qian and Z. Yin, Geotechnical Principles and Calculation, China Water Resource and Hydropower Press, Beijing, China, 2nd edition, 1994. .41
41. D. J. Henkel, "The shear strength of saturated remolded clays," in Research Conference on Shear Strength of Cohesive Soils, pp. 533–554, ASCE, Boulder, Colo, USA, 1960. .42
42. M. Chen, Y. Jin, and G. Zhang, Petroleum Related Rock Mechanics, Science Press, Beijing, China, 2008.

Early Diagenesis Records and Pore Water Composition of Methane-Seep Sediments from the Southeast Hainan Basin, South China Sea

Daidai Wu[1,2], Nengyou Wu[1,2], Ying Ye,[3] Mei Zhang[1,2], Lihua Liu[1,2], Hongxiang Guan[1,2], and Xiaorong Cong[1,2]

[1]Key Laboratory of Renewable Energy and Gas Hydrate, Guangzhou Institute of Energy Conversion, CAS, Guangzhou 510640, China

[2]Guangzhou Center for Gas Hydrate Research, CAS, Guangzhou 510640, China

[3]Department of Ocean Science and Engineering, Zhejiang University, Hangzhou 310028, China

ABSTRACT

Several authigenic minerals were identified by XRD and SEM analyses in shallow sediments from the Southeast Hainan Basin, on the northern slope of South China Sea. These minerals include miscellaneous carbonates, sulphates, and framboidal pyrite, and this mineral assemblage indicates the existence of gas hydrates and a methane seep. The assemblage and fabric features of the minerals are similar to those identified in cold-seep sediments, which are thought to be related to microorganisms fostered by dissolved methane. Chemical composition of pore water shows that the concentrations of SO_4^{2-}, Ca^{2+}, Mg^{2+}, and Sr^{2+} decrease clearly, and the ratios of Mg^{2+} to Ca^{2+} and Sr^{2+} to Ca^{2+} increase sharply with depth. These geochemical properties are similar to those where gas hydrates occur in the world. All results seem to indicate clearly the presence of gas hydrates or deep water oil (gas) reservoirs underneath the seafloor.

INTRODUCTION

Methane is a major component of cold fluids on the continental shelf and slope [1, 2]. It is also a major hydrocarbon gas source of deep oilgas and gas hydrates. Hydrocarbon gas can escape through the overlying marine sediment column, enter the water column, and form a coldseep [1, 3, 4]. Most of the hydrocarbon gas can generate authigenic carbonates at or under the seafloor via a series of biogeochemical reactions, and a small amount may even be transported into the water column or enter the atmosphere [5, 6]. Hydrocarbon gas seep provides nourishment for Archaea and sulfate- reducing bacteria (SRB) found in benthic sediments. These microorganisms consume sulfate and oxydize methane into HCO_3^-. This process causes changes of pore water chemical composition and results in the precipitation of carbonate [7, 8]. Methane seeps are an important geological phenomenon in marine sediment, and methane-rich cold seeps are assumed to be the preferential outcrop of oilgas and gas hydrate on the seafloor. Therefore, the knowledge of methane seeps is significant for oil-gas exploration. Furthermore, the greenhouse potential of methane is 24 times of carbon dioxide. Compared to the artificial carbon dioxide, the greenhouse effect of methane discharged by geodynamic and tectonic processes plays a primary role on the global change in the natural environment [9]. Therefore, there has been an increasing interest in methane seepage (gas hydrate) during the past several years.

Southeast Hainan Basin on the northern slope of South China Sea is a potential gas hydrate deposit. Except for direct drilling and sampling of marine gas hydrates, the occurrence of gas hydrates has been identified generally by inference from indirect evidence, derived from geological, geophysical, and geochemical data all over the world. For example, the occurrence of gas hydrate is inferred from seismic profiles, especially the bottom simulating reflector (BSR), although gas hydrates are also known to occur in the areas without BSR in many locations. In Southeast Hainan Basin, the high-resolution multichannel seismic profiles were collected, and BSR and other seismic indications, such as blanking zone and velocity amplitude anomaly inversions, for gas hydrate occurrences, were observed. However, geochemical studies have been of great help for identification of the occurrence of gas hydrates.

The goal of this research is to study the early diagenesis records and geochemistry of pore waters of shallow sediments from Southeast Hainan Basin. The evidences for gas hydrates or deep water oil (gas) reservoirs and the relationship between authigenic mineral, geochemistry of pore waters, and methane seep in shallow sediments were discussed here.

GEOLOGICAL SETTING AND SAMPLING

A 4.9 m long sediment core was taken in August 2005 from 1508 m depth in the Southeast Basin (Figure 1) by a gravity piston corer, during the HY4-2005^{-5} Cruise of R/V Haiyang- 4 (Guangzhou Marine Geological Survey, Ministry of Land and Natural Resources). The cored sediment is silty clay with a few foraminifera, yielding strong odor of hydrogen sulfide. The sediment core was stored on board at 4°C. The sampling site T1 (111°4' E, 18°2' N) is located in a flat continental slope, 150 km southeast of Sanya, Hainan Province, China. The geological structural unit belongs to the Songxi Depression Belt, Southeast Hainan Basin.

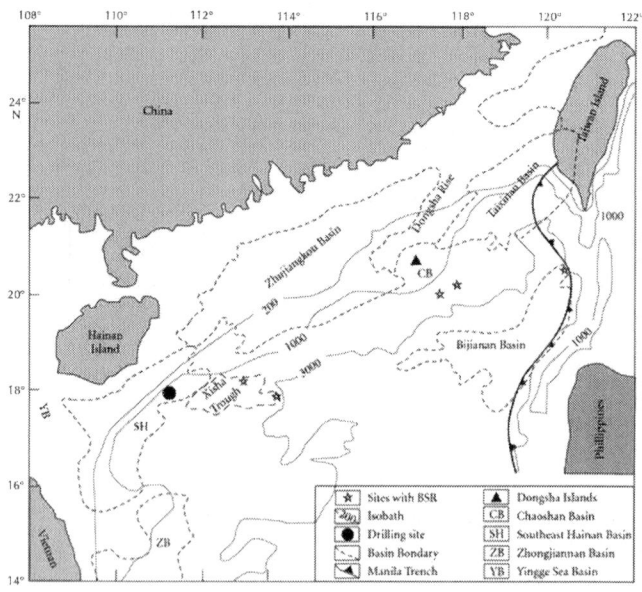

Figure 1: Position of the sampling site, Southeast Hainan Sea, South China Sea.

Oil- and gas-bearing depositional basins on the northern continental slope of the South China Sea were developed during the late Cenozoic [10–12], when a strong extrusion phase followed a period of rift extension and sea-floor spreading 30~24 Ma ago. The area (Figure 1) is characterized by frequent tectonic and magmatic episodes, slope slumps, high heat flux (average about 75 mW/m^2), abundant thermogenic gas and biogas in shallow sediments, carbon dioxide and nitrogen-rich gas. Four subareas, including Southwest Taiwan, Northeast Dongsha, Baiyun Sag of Pearl River Mouth Basin, and Southeast Hainan Basin, were identified to be the most favorable for gas seep or gas hydrate presence, showing geological, geochemical, and biological evidences for cold seeps [13–21]. The sediment thickness is reported to be about 12000 m [22], whereas 5000 m thick organic-rich sediments have been accumulated since the Cenozoic [23].

The thick Quaternary organic-rich sediments in Southeast Hainan Basin provide enough sources for hydrocarbons [24]. High geothermal gradient [25] and abnormal high pressure [26] are fitted for hydrocarbons formation, diffusion, and transportation. Methane-rich gas is categorized into three groups: bio-low mature transient gas, normal mature thermogenic gas, and super mature thermogenic gas [23], and the conditions for gas hydrate accumulation were confirmed in previous studies [27]. Some methane-seep-related features, such as mud volcanoes, mud diapirs, and gas chimneys, were discovered in the basin [24, 28]. Unfortunately, there is no similar report at the sampling site, even if the existence of gas hydrates in the Southeast Hainan Basin has been confirmed by geophysical evidences [24, 28–30]. The existence of gas hydrate in the same site (T1) has also been proved by microbial evidence [31].

MATERIALS AND METHODS

The sediment cores within PVC liners were immediately cut into 80 cm sections after its recovery on deck. The core sections were airtight sealed with plastic cap, gummed typed and stored on board at 4°C. The sealed sections were further splitted longitudinally in a flow-through (N_2) anaerobic chamber at Guangzhou Marine Geological Survey, China. One half of each section was described and archived, while the other half was sampled for pore water and sediment. The external rim of each subsection was removed using sterile tools to avoid contamination. Subsections of freshly cored sediments were removed and placed into special sealed glass bottles. The samples were stored at 4°C for X-ray diffraction (XRD), scanning electron microscope (SEM), and pore water analysis. All analyses were carried out at the Department of Geological Sciences and the Department of Biology, University of Miami.

XRD was employed to analyze mineralogy in the sediments. Samples were oven heated at 60°C for 12 hrs, pestled into less than 200 μ mesh, and X-rayed with a Scintag XGEN-4000 X-ray diffraction system, using Cu radiation in a graphite monochromator, a scintillation counter, with 0.02 two-theta steps from 5 to 70 degrees, for 2 seconds per step, with 0.15406 nm wavelength, 40 kV tube voltage, and 34 mA tube current. Mineral identification was done by a combination of search-match software and comparison with the JCPDS card files.

A Zeiss Supra 35 VP-FEG SEM equipped with EDAX Energy Dispersive X-ray microanalysis system was used to identify minerals and to image their morphologies. In preparation for SEM observations, selected samples were vacuum dried and gold coated. The working distance is 10.3 mm, and voltage is 5 kV.

Sediment pore water samples were collected after centrifugation of about 50 g sediment subsamples (5000 g for 10 minutes). Each sediment subsample could obtain about 13–20 mL water. To measure anions and cations, aliquots of pore water were mixed with 0.1 HCl solutions. The diluted (pore water samples) solutions were then analyzed by high-performance liquid chromatography (HPLC) for anions (IonPacAS14 column 4×250 mm) and direct current plasma emission spectrometry (DCP) for cations.

RESULTS AND DISCUSSIONS

Authigenic Minerals and Microstructures of Sediments

XRD results (Figure 2) show that the sediments contain land-derived detrital minerals, that is, quartz, illite, kaolinite, albite, calcite, and a complex suite of authigenic minerals, that is, miscellaneous carbonates, sulphates, and pyrite. Major mineral components and their semiquantification are listed in Table 1. Abundant microstructures were observed by SEM (Figure 3), and the whole crystal form, surface, angle, and crystallographic elements are kept intact during the observations. These characters confirm that the minerals are authigenic in situ, without transport and erosion.

Table 1: Major mineral components and their semiquantification results in sediments from the Southeast Hainan Basin (%)

No.	Depth (cm)	Illite	Gypsum	Kaolinite	Brucite	Witherite	Barite	Anhydrite	Quartz	Microcline	Albite	Ba-calcite	Calcite	Mg-calcite	Siderite	Magnesite	Vaterite	Pyrite	Maghemite
T1-44	20	24.2		11.9			22.3		13.1		2.3		7.9			3.5	3.4	11.5	
T1-46	35	25.4		13.4	7.0	1.3	22.5	3.5	12.0		2.7		8.9			3.5			
T1-46	85	28.7		14.2			25.9		13.8		2.6	5.8	6.0			3.0			
T1-36	100	25.2		9.1			30.2		14.6		2.5		8.7		1.9	3.3		4.5	
T1-38	200	30.3		11.7	9.6		25.0		10.4			4.4	3.7					4.9	
T1-28	216	30.8		13.4	3.8		28.0		10.7			5.4	5.9			3.7		8.3	
T1-30	230	26.4		13.4	10.0		28.9		9.1			3.0	5.3					3.9	
T1-32	320	27.4		16.2			28.2		14.5		4.0		9.6						
T1-20	336	24.5		10.7	11.6		27.1		10.5		2.3		3.7	2.3		4.6		5.1	
T1-22	355	19.7		11.1	10.4	16.0		2.5	13.4		1.6		11.7			3.1		3.3	
T1-24	380	24.41	3.08	11.37				2.94	26.50		3.27		25.16	2.51				0.76	5.0
T1-12	395	43.1		16.7	7.1		15.3		6.4		2.3		3.3	1.8				4.0	
T1-14	415			16.6			37.8	6.6	13.5		6.2		7.6	2.7				9.0	
T1-16	440	20.8		17.9				2.8	29.4		3.4		19.9	3.1		2.7			
T1-2	450	26.1	4.3	15.0					26.4		3.6		19.0	2.9		2.6			
T1-4	465	20.3		14.8		1.7			30.0	9.9	3.3		18.2					1.8	
T1-6	475	22.1		11.3				3.2	29.8	8.0	3.8		15.1	2.3		2.6		1.7	

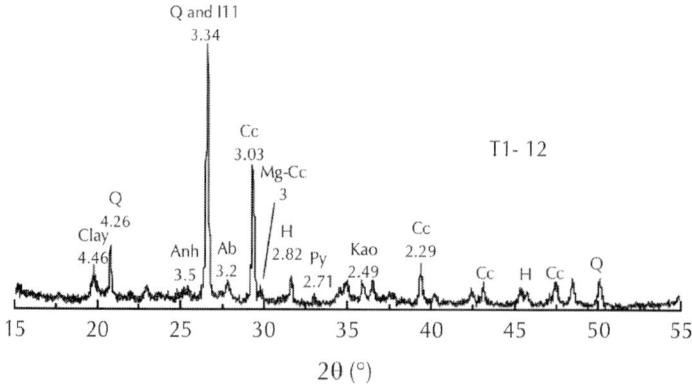

Figure 2: XRD pattern of mineral components in sample T1-12 from the Southeast Hainan Basin. Clay, clay minerals; Q, quartz; Anh, anhydrite; Ill, illite; Ab, albite; Cc, calcite; Mg-Cc, Mg-calcite; H, halite; Py, pyrite; Kao, kaolinite.

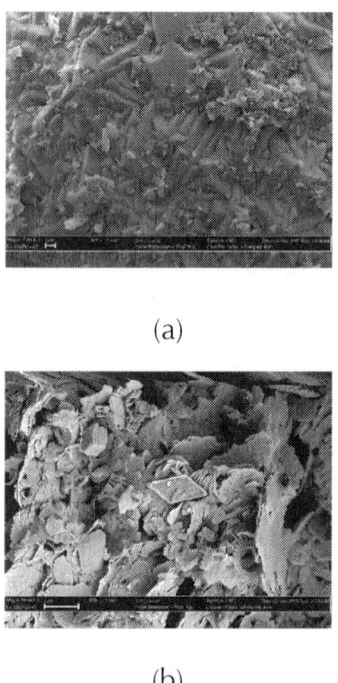

(a)

(b)

Early Diagenesis Records and Pore Water Composition... 31

(c)

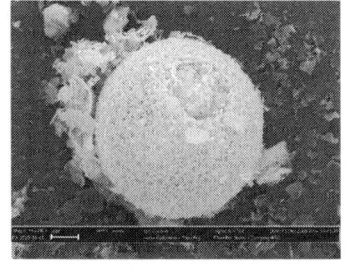

(d)

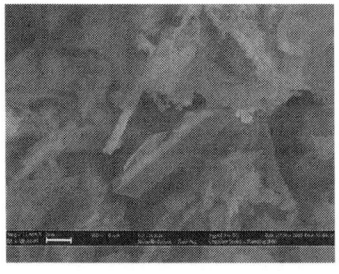

(e)

(f)

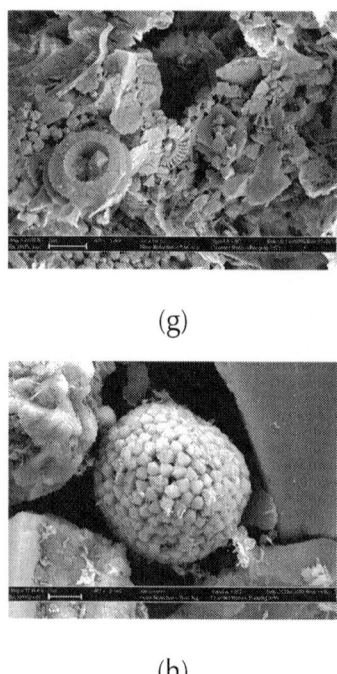

(g)

(h)

Figure 3: SEM images of authigenic minerals in sediments from the Southeast Hainan Basin.

Main identified miscellaneous carbonates are calcite, Mg-calcite, nesquehonite, magnesite, siderite, and so forth. Calcite is one of the major components in the sediment samples, and most part of it is foraminifera shell. Mg-calcite and nesquehonite are common in XRD patterns, but magnesite and siderite are only identified in few subsamples. Part of calcite is aragonite pseudomorph, which looks like an assemblage with acicular, filiform, and ball-shaped carbonates under SEM. The assemblage is identified with calcite by energy dispersive X-ray spectra. However, the crystal form and habit are similar to aragonite (Figures 3(c), 3(d)). There is no identified aragonite peaks in the XRD pattern, which suggests that aragonite was replaced by calcite, and the pseudomorph of aragonite is observed. It evinces that phase transformation of aragonite into calcite has occurred in sediments. The factor leading to formation of acicular, filiform, and ball-shaped aragonite is metabolism of methane-related microorganisms. Sassen et al. found that acicular aragonite, framboidal pyrites, mycelium, and asphalt have symbiotic relationship in cold seep carbonates [5], and the bacterium activity boosting precipitation of authigenic carbonates has been discovered by lab work [32, 33]. Ehrlich observed successfully ball-shaped and dumbbell assemblage of acicular aragonite by bacilli culture in laboratory [34]. Ball-shaped carbonates cemented by mycelium were observed under SEM (Figure 3(d)), which further confirmed that there was some cause-effect relationship between carbonates assemblage and microorganism. Mg-calcite and Nesquehonite are authigenic rhombohedral crystals, and usually coat the surface of microorganism shells like coccoliths, diatoms, and so on [5].

Authigenic carbonates deposits have been documented in area near methane seep in both the active and passive continental margin [5, 6]. The carbonates formation relates closely to methane seep or gas hydrate decomposition because of their special microstructure [2, 35]. The Southeast Hainan Basin in South China Sea is a potential area for gas hydrate reservoirs. Methane microseep and its oxidation drive the bicarbonate deposited associating with the available research data although there are no large methane seep found at the sampling site.

Main sulphate minerals identified in the sediments are barite, anhydrate, and gypsum. Gypsum and anhydrite appear as tabular and panidiomorphic textures (Figure 3(e)). The occurrence of authigenic gypsum and anhydrite indicates high concentration of sulphate ion in the pore water. Sulfate reduction is the main pathway of CH4 oxidation in CH4-riched anoxic sediments. Although the CH_4-SO_4^{2-} redox cell will decompose and consume sulphate in the sediments, that does not mean there is no SO4 2−. On the contrary, SO4 2− originates from the decomposition of organic-rich sediments, and there must be abundant organic matter in gas hydrate reservoirs [36], whereas Ca2+ may originate from the decomposition of calcilith or fluids with Ca^{2+} from the gas hydrate present underneath [37]. Therefore, the supersaturated concentration of Ca^{2+}, and SO_4^{2-} in pore water drives authigenic gypsum precipitation. Authigenic gypsum has been identified in the sediments from Hydrate Ridge in the northeast Pacific Ocean. That indicates gypsum should be one of the authigenic minerals which have close relation with methane hydrate reservoir [38]. In conclusion, gypsum is one of the markers for hydrate presence similar to other authigenic minerals, carbonates, and barite.

Barite shows a short prismatic and euhedral crystal structure under SEM (Figure 3(f)). Authigenic barite micro-crystals had been indentified in sediments from the Peru continental margin (site 684) and Japan Sea (site 799) by the Ocean Drilling Program (ODP), and they contain more 34S than that in the sea water with a $\delta^{34}S$ of +84‰. The widespread occurrence of barite indicates high concentration of Ba in pore water. Barite is the major mineral in the coldseep sediments in Peru and Russian Okhotsk [39, 40]. The source of Ba in the cold seep is assumed to be the barite solution as a result of the sulphate deoxidization at depth [41].

Pyrite was identified in most samples by XRD analysis, and the representative diffraction peaks are 2.709×10^{-10} and 2.423×10^{-10} m. Pyrite is distributed as single granules (Figure 3(g)) or framboidal assemblages (Figure 3(h)); $\delta^{34}S$ value of the framboidal pyrite is greatly negative in coldseep sediments and low-temperature hydrothermal fluid sediments. This indicates that the sulphur of pyrite originates from bacterial reduction of sulphate in sea water [42–44]. Sulphate bacteria mainly depend on sulphate decomposition in the anoxic environment by activity of sulphate redox cell. Meanwhile, the HCO_3^- concentration increases in the pore water with decomposition of organic matter. Sulphide forms with free HS^- and low-valence iron, and finally sulphide transforms into pyrite [45, 46]. Sassen et al. found many framboidal pyrites, and pyrite granules in sediments from a gas hydrate area in the Golf of Mexico and suggested that the strong anaerobic methane oxidation (AOM) contributes to pyrite formation in sediments [5]. Liu et al. also believed that AOM is the main reason for sulphide increase in the sediments from the northeastern South China Sea [47]. Therefore, authigenic pyrite is also the major product of AOM process in sediments from a gas hydrate (methane seep) area. High concentration of pyrite in anoxic sediments can indicate the methane abnormality [48] and the occurrence of methane-rich fluids and further show the shallow SMI to a certain degree.

In summary, AOM in the methane seep area results in alkalinity increase in both pore water and sea water [49,50]. Methane is oxidized to HCO_3^- by SO_4^{2-}), a process in favor of CO_3^{2-} accumulation. Furthermore, AOM changes the geochemical conditions and results in the rapid deposition of miscellaneous carbonates including aragonite, Mg-calcite, and dolomite [51]. Meanwhile, pyrite formation results from the reaction of H_2S, derived from bacterial sulphate reduction, with reactive iron [45, 46]. Precipitation of aragonite, Mg-calcite, and pyrite is the result of processes related to the presence of methane seep in the northern Black sea [1, 8]. Abundant cold-seep-related minerals, Mg-calcite, dolomite, and framboidal pyrite, precipitate in Monterey Gulf, California [52]. Authigenic minerals including miscellaneous carbonates, sulphate, and sulphide form in gas hydrate-bearing and hydrocarbon seepage in Gulf of Mexico [5]. All these records of authigenic mineral assemblages and fabric features are geological evidences for cool-seep activities, hydrocarbon seeps, and gas hydrate dissociation. The widespread occurrence of authigenic carbonates, pyrite, and sulphate in the sediments of Southeast Hainan Basin, South China Sea is completely similar to reported cases in a typical gas hydrate area. This aspect further indicates gas hydrate existence in the study area.

Chemical Composition of Pore Waters

Anomalies in the chemical composition of pore water indicating the occurrence of gas hydrates, such as anomalous concentrations of Br, Cl, CH_4, and K, are known in the literature [53, 54].

In pore waters from the Southeast Hainan Basin, NH_4^+ concentration is close to zero within the upper 320 cm section and increases with depth in lower section (Table 2, Figure 4). That may be due to the zymosis of organic matter via bacteria during gas hydrate formation and dissociation [55]. The concentration of Mg^{2+} decreases slightly with depth, and the Ca^{2+} concentration falls slightly (Table 2, Figure 4). The solubility product of calcium carbonate is less than magnesium, and calcium carbonate deposits easily from the solution. Therefore, Ca^{2+} deposits faster than Mg^{2+} from pore water into solid phase. Furthermore, the Mg2+/Ca2+ ratio increases sharply with depth (Figure 4). The concentrations of Sr^{2+}, and Mn^{2+} go down sharply with depth in the shallow sediments (Table 2). This suggests that Sr^{2+} and Mn^{2+} are removed into the same authigenic phase(s). Also the ratio of Sr_{2+} to Ca^{2+} increases sharply with depth (Figure 4), and the cause may be that Ca^{2+}, Mg^{2+}, Sr^{2+}, and Mn^{2+} cations are easily incorporated within carbonates deposited at the expense of dissolved carbon dioxide. The consumption of Ca^{2+}, Mg^{2+}, Sr^{2+}, and Mn^{2+} cations in pore water, combined with the authigenic carbonates like calcite (aragonite pseudomorphous), Mg-calcite, and magnesite, indicates that there is supplement of dissolved CO2 likely originated from AOM in depth during the diagenesis process, then diagenesis is a process affecting AOM and providing cations and dissolved species to pore water for further reactions. Previous research showed that carbon dioxide could form during gas hydrate decomposition and upper migration in gas hydrate-bearing areas, leading to formation of authigenic carbonates, syngenetic to gas hydrate [56, 57].

Table 2: Pore water chemical composition for the analysed core from the Southeast Hainan Basin (mg/dm^3)

Depth (cm)	NH_4	K^+	Mg^{2+}	Ca^{2+}	Mn^{2+}	Fe^{2+}	F^-	Br^-	SO_4^{2-}
20	0.0	585.1	1465.9	510.1	17.7	28.8	3.0	71.4	2559.5
35	0.0	561.4	1421.1	529.1	17.2	19.9	2.7	69.3	2496.8
85	0.0	588.1	1448.6	487.3	16.8	45.6	3.0	70.8	2447.2
100	0.0	571.2	1393.3	466.7	15.8	36.6	4.4	69.7	2308.3
200	0.0	592.1	1446.4	469.2	9.0	35.7	4.2	79.1	2423.1
216	0.0	558.9	1342.0	429.9	8.5	30.5	3.2	65.3	2199.2
230	0.0	592.8	1412.7	461.5	8.8	48.7	3.0	68.1	2323.2
320	0.0	674.6	1521.3	431.2	7.5	18.1	4.1	71.5	2014.3
336	10.7	558.5	1412.9	439.9	8.0	27.0	3.5	75.4	2175.8
355	0.0	565.9	1460.6	466.5	8.8	46.9	3.1	78.5	2199.9
380	26.4	552.7	1424.1	429.8	7.8	18.0	3.0	67.8	2170.3
395	22.8	595.4	1415.8	423.2	7.9	28.6	3.2	76.0	2173.7
415	15.7	546.4	1406.7	439.1	8.9	53.2	3.7	71.8	2074.6
440	44.7	537.8	1375.7	412.3	12.0	22.2	3.0	72.7	2085.0
450	55.5	579.6	1421.8	412.7	7.6	9.0	3.8	77.6	2163.3
465	53.6	550.5	1434.4	425.2	8.0	14.4	4.7	79.2	2078.0
475	61.3	595.2	1407.1	363.0	7.3	25.2	4.1	72.6	2153.7

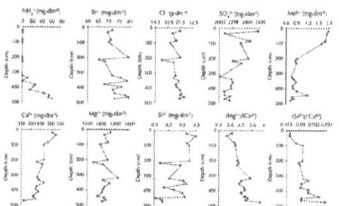

Figure 4: Geochemistry of pore waters versus depth of sediments from the Southeast Hainan Basin.

The concentrations of Cl^- and Br^- are relatively constant, whereas SO_4^{2-} decreases clearly with depth (Figure 4). These trends, combined with the presence of authigenic carbonates, barite, gypsum, anhydrite, and pyrite, indicate AOM and sulphate reduction

$$CH_4 + SO_4^{2-} \longrightarrow HCO_3^- + HS^- + H_2O \qquad (1)$$

The reaction has affected the shallow sediments at the site T1 of the Southeast Hainan Basin, South China Sea, or the sulphate-methane interface (SMI) has been lifted up to the shallow layer. It means that there is abundance of gas hydrate or gas-oil resource underneath the sampling site. Compared to other methane seep area, the concentration gradient of sulphate in core T1 is smaller. That means sulfate in pore water is not exhausted in sampling depth. So, the depth of SMI should be deeper than the sampling's. Therefore, it is clear that there is a hypogenetic fluid, related to gas hydrate occurrence under the sampling site.

The results of microbial colony research at same site [31] demonstrated that physiological functions of these relatives include Fe(III) and Mn(IV) reduction (Pelobacter), sulfate reduction (H2S production) (S. kaireitica), decomposition of complex aromatic hydrocarbons and denitrification (P. stutzeri), hydrocarbon degradation (A. jadensis), and thiosulfate/sulfite oxidation (Sulfitobacter spp.) in the sediments, similarly to observations of gas hydrate-bearing sediments in the Sea of Okhotsk [58]. The occurrence of microbe suggests hydrocarbon oxidation coupled with sulphate, and Fe (III) reduction is an important process during the early diagenesis in the Southeast Hainan Basin, South China Sea. Oxidation of hydrocarbons may contribute to the increasing alkalinity and carbonate precipitation. The metabolic processes are consistent with the decreasing Mg^{2+}, Ca^{2+}, Sr^{2+} and SO_4^{2-} concentrations in pore water, the precipitation of authigenic carbonates and pyrite in the shallow sediments. These results strongly suggest that there should be gas hydrae reservoirs in the sampling area.

CONCLUSIONS

Taking together all available mineralogical and geochemical data from the Southeast Hainan Basin, South China Sea, we can draw the following conclusions.(1) The shallow sediments from the Southeast Hainan Basin are of a complex suite of authigenic minerals, such as carbonates, sulphates, and framboidal pyrite, indicating clearly the existence of gas hydrates or deep water oil (gas) reservoirs. The assemblage and fabric features of aragonite and mg-calcite could be a result of the activity of microorganisms, which consumes dissolved methane. The miscellaneous authigenic carbonates are the result of abundant HCO_3^- in the pore water, whose source should be related to AOM by microorganisms. Reduction of sulphate and iron may be coupled with oxidation of methane and hydrocarbons, which causes decreased SO_4^{2-} concentration in pore water and precipitation of authigenic pyrite and carbonates in the sediments.

(2) Chemical composition of pore waters shows that the concentrations of SO_4^{2-}, Mg^{2+}, Ca^{2+}, Sr^{2+} and Mn^{2+} decrease, and the concentration of NH_4^+ increases with depth. The ratios of Mg^{2+}/Ca^{2+}, Sr^{2+}/Ca^{2+} increase sharply with depth. These geochemical data indicate strongly the presence of gas hydrates or deep water oil (gas) reservoirs beneath the seafloor.

In summary, mineralogical and geochemical data in the shallow sediments and pore waters suggest to us that the Southeast Hainan Basin is one of the promising targets on the northern margin of the South China Sea for further gas hydrate or deep water oil (gas) reservoir exploration. The data allow us to better understand the influence on the methane seep and geochemistry of fluids in shallow sediments exerted by gas hydrates or deep water oil (gas) reservoirs, whose presence could not be determined with certainty until there is further evidence, and the presence of methane seeps on the seafloor needs to be evidenced by swath bathymetry, shallow seismic data, or direct evidence on seafloor images in the future.

ACKNOWLEDGMENTS

This study is financially supported by the National Natural Science Foundation of China (No. 41003010 and No.U0933004), National 973 Project (No. 2009CB219506), and Scientific and Technology Program of Guangdong Province (No. 2011A080403021). We sincerely thank the crew and scientific team onboard R/V Haiyang-4 from Guangzhou Marine Geological Survey for their cooperation and highly professional support, two anonymous reviewers for their kind comments and suggestions.

REFERENCES

1. J. Peckmann, A. Reimer, U. Luth et al., "Methane-derived carbonates and authigenic pyrite from the northwestern Black Sea," Marine Geology, vol. 177, no. 1-2, pp. 129–150, 2001.
2. C. Pierre and J. M. Rouchy, "Isotopic compositions of diagenetic dolomites in the Tortonian marls of the western Mediterranean margins: evidence of past gas hydrate formation and dissociation," Chemical Geology, vol. 205, no. 3-4, pp. 469– 484, 2004.
3. J. Peckmann and V. Thiel, "Carbon cycling at ancient methane-seeps," Chemical Geology, vol. 205, no. 3-4, pp. 443–467, 2004.
4. D. Wu, N. Wu, Y. Ye, P. Zhang, and X. Chen, "Geochemical characteristics of hydrocarbon compounds in sediments of the eastern South China Sea," Acta Petrolei Sinica, vol. 29, no. 4, pp. 516–521, 526, 2008 (Chinese).
5. R. Sassen, H. H. Roberts, R. Carney et al., "Free hydrocarbon gas, gas hydrate, and authigenic minerals in chemosynthetic communities of the northern Gulf of Mexico continental slope: relation to microbial processes," Chemical Geology, vol. 205, no. 3-4, pp. 195–217, 2004. Journal of Geological Research 9
6. B. M. A. Teichert, G. Bohrmann, and E. Suess, "Chemoherms on Hydrate Ridge—unique microbially-mediated carbonate build-ups growing into the water column," Palaeogeography, Palaeoclimatology, Palaeoecology, vol. 227, no. 1–3, pp. 67–85, 2005.
7. C. L. Zhang and B. Lanoil, "Geological and biogeochemical dynamics of gas hydrate-hydrocarbon seep systems," Chemical Geology, vol. 205, pp. 187–194, 2004.
8. J. Reitner, J. Peckmann, M. Blumenberg, W. Michaelis, A. Reimer, and V. Thiel, "Concretionary methane-seep carbonates and associated microbial communities in Black Sea sediments," Palaeogeography, Palaeoclimatology, Palaeoecology, vol. 227,

no. 1–3, pp. 18–30, 2005.

9. G. Etiope, "New directions: GEM—geologic emissions of methane, the missing source in the atmospheric methane budget," Atmospheric Environment, vol. 38, no. 19, pp. 3099–3100, 2004.

10. S. L. McDonnell and M. Czarnecki, "A note on gas hydrate in the northern sector of the South China Sea," in Natural Gas Hydrate: In Oceanic and Permafrost Environments, M. D. Max, Ed., pp. 239–244, Kluwer Academic Publishers, Dodrecht, The Netherlands, 2000.

11. T. M. Guo, B. H. Wu, Y. H. Zhu, S. S. Fan, and G. J. Chen, "A review on the gas hydrate research in China," Journal of Petroleum Science and Engineering, vol. 41, no. 1–3, pp. 11–20, 2004.

12. S. Wu, G. Zhang, Y. Huang, J. Liang, and H. K. Wong, "Gas hydrate occurrence on the continental slope of the northern South China Sea," Marine and Petroleum Geology, vol. 22, no. 3, pp. 403–412, 2005.

13. D.-F. Chen, Y.-Y. Huang, D. Feng, Z. Su, and G.-Q. Chen, "Seep carbonate and preserved bacteria fossils in the northern of the South China Sea and their geological implications," Bulletin of Mineralogy Petrology and Geochemistry, vol. 24, no. 3, pp. 185–189, 2005 (Chinese).

14. H. Lu, J. Liu, F. Chen et al., "Mineralogy and stable isotopic composition of authigenic carbonates in bottom sediments in the offshore area of southwest Taiwan, South China Sea: evidence for gas hydrate occurrence," Earth Science Frontiers, vol. 12, no. 3, pp. 268–276, 2005 (Chinese).

15. Z. Chen, W. Yan, M. Chen et al., "Discovery of seep carbonate nodules as new evidence for gas venting on the northern continental slope of South China Sea," Chinese Science Bulletin, vol. 51, no. 10, pp. 1228–1237, 2006.

16. D. F. Chen, Y. Y. Huang, X. L. Yuan, and L. M. Cathles, "Seep carbonates and preserved methane oxidizing archaea and sulfate reducing bacteria fossils suggest recent gas venting on the seafloor in the Northeastern South China Sea," Marine and Petroleum Geology, vol. 22, no. 5, pp. 613–621, 2005.

17. S. Lin, W. C. Hsieh, Y. C. Lim, T. F. Yang, C. S. Liu, and Y. Wang,

"Methane migration and its influence on sulfate reduction in the good weather ridge region, South China Sea continental margin sediments," Terrestrial, Atmospheric and Oceanic Sciences, vol. 17, no. 4, pp. 883–902, 2006.

18. T. F. Yang, P. C. Chuang, S. Lin, J. C. Chen, Y. Wang, and S. H. Chung, "Methane venting in gas hydrate potential area off-shore of SW Taiwan: evidence of gas analysis of water column samples," Terrestrial, Atmospheric and Oceanic Sciences, vol. 17, no. 4, pp. 933–950, 2006.

19. X. Su, F. Chen, S. Wei et al., "Preliminary study on the correlation between microbial abundance and methane concentration in sediments from cold seeps in the northern South China Sea," Geoscience, vol. 21, no. 1, pp. 101–104, 2007 (Chinese).

20. Y. Y. Huang, E. Suess, N. Y. Wu et al., "Methane and gas hydrate geology of the Northern South China Sea—Sino-German Cooperative," Cruise report SO-177, Geological Publishing House, Bejingm, China, 2008.

21. X. Han, E. Suess, Y. Huang et al., "Jiulong methane reef: microbial mediation of seep carbonates in the South China Sea," Marine Geology, vol. 249, no. 3-4, pp. 243–256, 2008.

22. B. Huang, "Gas Potential and its favorable exploration areas in Southeast Hainan Basin," Natural Gas Industry, vol. 19, no. 1, pp. 34–40, 1999 (Chinese).

23. J. He, B. Xia, D. Sun et al., "Hydrocarbon accumulation, migration and play targets in the Southeast Hainan Basin, South China Sea," Petroleum Exploration and Development, vol. 33, no. 1, pp. 53–58, 2006 (Chinese).

24. B. Wu, G. Zhang, Y. Zhu et al., "Progress of gas hydrate investigation in China offshore," Earth Science Frontiers, vol. 10, no. 1, pp. 177–188, 2003 (Chinese).

25. J. Liu and C. Wang, "Thermal fluid in Ying-Qiong Basin and its significance of oil-gas geology," Natural Gas Exploration and Development, vol. 27, no. 1, pp. 12–15, 2004 (Chinese).

26. M. Wang, "The character of overpressure and its relationship with the distribution of oil and gas Southeast Hainan Basin," Offshore Oil, vol. 23, no. 1, pp. 15–21, 2003 (Chinese).

27. J. He, "The evolving of gas hydrate and the exploration foreground

in the north of South China Sea," Offshore Oil, vol. 23, no. 1, pp. 57–64, 2003 (Chinese).
28. D. Chen, X. Li, and B. Xia, "Distribution of gas hydrate stable zones and resource prediction in the Southeast Hainan Basin of the South China Sea," Chinese Journal of Geophysics, vol. 47, pp. 483–489, 2004 (Chinese).
29. S. Jiang, T. Yang, Z. Xue et al., "Chlorine and sulfate concentrations in pore waters from marine sediments in the north margin of the South China Sea and their implications for gas hydrate exploration," Geoscience, vol. 19, pp. 45–54, 2005 (Chinese).
30. X. Su, F. Chen, X. Yu et al., "A pilot study on Miocene through Holocene sediments from the continental slope of the South China Sea in correlation with possible distribution of gas hydrates," Geoscience, vol. 19, pp. 1–3, 2005 (Chinese).
31. H. Jiang, H. Dong, S. Ji, Y. Ye, and N. Wu, "Microbila diversity in the deep marine sediment from the Qiongdongnan Basin in China Sea," Geomicrobiology Journal, vol. 24, no. 6, pp. 505–517, 2007.
32. Y. Van Lith, R. Warthmann, C. Vasconcelos, and J. A. McKenzie, "Microbial fossilization in carbonate sediments: a result of the bacterial surface involvement in dolomite precipitation," Sedimentology, vol. 50, no. 2, pp. 237–245, 2003.
33. D. T. Wright and D. Wacey, "Precipitation of dolomite using sulphate-reducing bacteria from the Coorong Region, South Australia: significance and implications," Sedimentology, vol. 52, no. 5, pp. 987–1008, 2005.
34. H. L. Ehrlich, "Microbial formation and degrafation of carbonates," in Geomicrobiology, pp. 183–228, Marcel Dekker, New York, NY, USA, 4th edition, 2002.
35. S. Cavagna, P. Clari, and L. Martire, "The role of bacteria in the formation of cold seep carbonates: geological evidence from Monferrato (Tertiary, NW Italy)," Sedimentary Geology, vol. 126, no. 1–4, pp. 253–270, 1999.
36. M. V. S. Guptha, "Authigenic gypsum in a deep sea core from southeastern Arabian sea," Journal of the Geological Society of India, vol. 21, no. 11, pp. 568–571, 1980.
37. E. Suess, M. E. Torres, G. Bohrmann et al., "Gas hydrate

destabilization: enhanced dewatering, benthic material turnover and large methane plumes at the Cascadia convergent margin," 10 Journal of Geological Research Earth and Planetary Science Letters, vol. 170, no. 1-2, pp. 1–15, 1999.
38. J. Wang, E. Suess, and D. Rickert, "Authigenic gypsum found in gas hydrate-associated sediments from Hydrate Ridge, the eastern north Pacific," Scinece in China D, vol. 33, no. 5, pp. 433–441, 2003 (Chinese).
39. M. E. Torres, H. J. Brumsack, G. Bohrmann, and K. C. Emeis, "Barite fronts in continental margin sediments: a new look at barium remobilization in the zone of sulfate reduction and formation of heavy barites in diagenetic fronts," Chemical Geology, vol. 127, no. 1-3, pp. 125–139, 1996.
40. J. Greinert, G. Bohrmann, and E. Suess, "Gas hydrate-associated carbonates and methane-venting at Hydrate Ridge: classification, distribution and origin of authigenic lithologies," in Natural Gas Hydrates: Occurrence, Distribution and Detection, C. K. Paull and W. P. Dillon, Eds., vol. 124, pp. 99–113, American Geophysical Union, Washington, DC, USA, 2001.
41. G. Aloisi, K. Wallmann, S. M. Bollwerk, A. Derkachev, G. Bohrmann, and E. Suess, "The effect of dissolved barium on biogeochemical processes at cold seeps," Geochimica et Cosmochimica Acta, vol. 68, no. 8, pp. 1735–1748, 2004.
42. M. E. Bottcher and A. Lepland, "Biogeochemistry of sulfur in a ¨ sediment core from the west-central Baltic Sea: evidence from stable isotopes and pyrite textures," Journal of Marine Systems, vol. 25, no. 3-4, pp. 299–312, 2000.
43. R. T. Wilkin and M. A. Arthur, "Variations in pyrite texture, sulfur isotope composition, and iron systematics in the black sea: evidence for late pleistocene to holocene excursions of the O2-H2S redox transition," Geochimica et Cosmochimica Acta, vol. 65, no. 9, pp. 1399–1416, 2001.
44. P. Alfonso, R. M. Prol-Ledesma, C. Canet, J. C. Melgarejo, and A. E. Fallick, "Sulfur isotope geochemistry of the submarine hydrothermal coastal vents of Punta Mita, Mexico," Journal of Geochemical Exploration, vol. 78-79, pp. 301–304, 2003.
45. S. Lin, K. M. Huang, and S. K. Chen, "Organic carbon deposition and its control on iron sulfide formation of the southern East China

Sea continental shelf sediments," Continental Shelf Research, vol. 20, no. 4-5, pp. 619–635, 2000.
46. M. A. A. Schoonen, "Mechanisms of sedimentary pyrite formation," in Sulfur Biogeochemistry: Past and Present, J. P. Amend, K. J. Edwards, and T. W. Lyons, Eds., pp. 117–134, The Geological Society of America, Colo, USA, 2004.
47. J. Liu, H. Lu, Z. Liao et al., "Distribution in sulfides in shallow sediments in Dongsha area, South China Sea, and its relationship to gas hydrates," Earth Science Froniers, vol. 12, no. 3, pp. 258–262, 2005 (Chinese).
48. H. Lu, F. Chen, J. Liu et al., "Authigenic mineral associated with sedimentary environment of gas hydrate deposit and their occurrence in South China Sea," Geological Research of South China Sea, vol. 1, pp. 93–104, 2006 (Chinese).
49. S. J. Mazzullo, "Organogenic dolomitization in peritidal to deep-sea sediments," Journal of Sedimentary Research, vol. 70, no. 1, pp. 10–23, 2000.
50. J. W. Pohlman, C. Ruppel, D. R. Hutchinson, R. Downer, and R. B. Coffin, "Assessing sulfate reduction and methane cycling in a high salinity pore water system in the northern Gulf of Mexico," Marine and Petroleum Geology, vol. 25, no. 9, pp. 942–951, 2008.
51. S. Ritger, B. Carson, and E. Suess, "Methane-derived authigenic carbonates formed by subduction- induced pore-water expulsion along the Oregon/Washington margin," Geological Society of America Bulletin, vol. 98, no. 2, pp. 147–156, 1987.
52. D. S. Stakes, D. Orange, J. B. Paduan, K. A. Salamy, and N. Maher, "Cold-seeps and authigenic carbonate formation in Monterey Bay, California," Marine Geology, vol. 159, no. 1–4, pp. 93–109, 1999.
53. W. S. Borowski, C. K. Paull, and W. Ussler, "Global and local variations of interstitial sulfate gradients in deep-water, continental margin sediments: sensitivity to underlying methane and gas hydrates," Marine Geology, vol. 159, no. 1–4, pp. 131–154, 1999.
54. S. P. Hesselbo, D. R. Grocke, H. C. Jenkyns et al., "Massive ¨ dissociation of gas hydrate during a Jurassic oceanic anoxic

event," Nature, vol. 406, no. 6794, pp. 392–395, 2000.

55. C. K. Paull, R. Matsumoto, P. J. Wallace et al., Proceedings of the Ocean Drilling Program, vol. 164, National Science Foundation and Joint Oceanographic Institutions, College Station, Tex, USA, 1996.

56. S. J. Burns, "Early diagenesis in Amazon Fan sediments," in Proceedings of the Ocean Drilling Program, Scientific Results, R. D. Flood, D. J. W. Piper, A. Klaus et al., Eds., vol. 155, pp. 497–504, National Science Foundation and Joint Oceanographic Institutions, College Station, Tex, USA, 1997.

57. N. M. Rodriguez, C. K. Paull, and W. S. Borowski, "Zonation of authigenic carbonates within gas hydrate-bearing sedimentary sections on the Blake Ridge: offshore southeastern North America," in Proceedings of the Ocean Drilling Program, Scientific Results, C. K. Paull, R. Matsumoto, P. J. Wallace et al., Eds., vol. 164, pp. 301–312, National Science Foundation and Joint Oceanographic Institutions, College Station, Tex, USA, 2000.

58. F. Inagaki, M. Suzuki, K. Takai et al., "Microbial communities associated with geological horizons in coastal subseafloor sediments from the Sea of Okhotsk," Applied and Environmental Microbiology, vol. 69, no. 12, pp. 7224–7235, 2003.

Chapter 3

Assessment of the Deepwater Horizon Oil Spill Impact on Gulf Coast Microbial Communities

Regina Lamendella[1,2], Steven Strutt[2], Sharon Borglin[1], Romy Chakraborty[1], Neslihan Tas[1], Olivia U. Mason[1,3], Jenni Hultman[1,4], Emmanuel Prestat[1], Terry C. Hazen[1,5,6], and Janet K. Jansson[1,7]

[1]Lawrence Berkeley National Laboratory, Earth Sciences Division, Ecology Department, Berkeley, CA, USA

[2]Biology Department, Juniata College, Huntingdon, PA, USA

[3]Department of Earth, Ocean and Atmospheric Science, Florida State University, Tallahassee, FL, USA

[4]Department of Food Hygiene and Environmental Health, University of Helsinki, Helsinki, Finland

[5]Department of Civil and Environmental Engineering, University of Tennessee, Knoxville, TN, USA

[6]Oak Ridge National Laboratory, Biosciences Division, Oak Ridge, TN, USA

[7]Department of Energy, Joint Genome Institute, Walnut Creek, CA, USA

ABSTRACT

One of the major environmental concerns of the Deepwater Horizon oil spill in the Gulf of Mexico was the ecological impact of the oil that reached shorelines of the Gulf Coast. Here we investigated the impact of the oil on the microbial composition in beach samples collected in June 2010 along a heavily impacted shoreline near Grand Isle, Louisiana. Successional changes in the microbial community structure due to the oil contamination were determined by deep sequencing of 16S rRNA genes. Metatranscriptomics was used to determine expression of functional genes involved in hydrocarbon degradation processes. In addition, potential hydrocarbon-degrading Bacteria were obtained in culture. The 16S data revealed that highly contaminated samples had higher abundances of Alpha- and Gammaproteobacteria sequences. Successional changes in these classes were observed over time, during which the oil was partially degraded. The metatranscriptome data revealed that PAH, n-alkane, and toluene degradation genes were expressed in the contaminated samples, with high homology to genes from Alteromonadales, Rhodobacterales, andPseudomonales. Notably, Marinobacter (Gammaproteobacteria) had the highest representation of expressed genes in the samples. A Marinobacter isolated from this beach was shown to have potential for transformation of hydrocarbons in incubation experiments with oil obtained from the Mississippi Canyon Block 252 (MC252) well; collected during the Deepwater Horizon spill. The combined data revealed a response of the beach microbial community to oil contaminants, including prevalence of Bacteria endowed with the functional capacity to degrade oil.

INTRODUCTION

The Deepwater Horizon oil spill was the largest accidental marine oil spill in the history of the oil industry, spewing an estimated 4.1 million barrels of crude oil into the Gulf of Mexico (Zukunft, 2010).

In addition, 1.84 million gallons of chemical dispersants were applied to assist in oil dispersal (The Federal Intragency Solutions Group: Oil Budget Calculator Science and Engineering Team, 2010). Physical barriers, direct collection from the wellhead, skimming, and burning were also implemented in order to mitigate the effects of the spill. Despite significant efforts to protect hundreds of miles of beaches, wetlands, and estuaries from oil, it began washing up on the Gulf Coast by early May 2010 (OSAT-2, 2011). Most recent estimates indicate that up to 22% of the 4.1 million barrels of oil was either trapped under the surface of the water as sheen, carried on the water surface as conglomerated tar (Lubchenco et al., 2010; Kimes et al., 2013), or deposited onto surface sediments (US Coast Guard, USGS, and NOAA, 2010; Mason et al., 2004). Some of the oil washed ashore where it was either collected or became entrained in sand and sediments. The contamination of beach ecosystems raised considerable concern due to the potential for detrimental environmental and economic impacts in the Gulf region (McCrea-Strub et al., 2011; Sumaila et al., 2012).

Initial research studies of the Gulf oil spill mainly focused on the fate of the oil in the water column. These studies highlighted the significant contribution of microorganisms toward the degradation of oil in a deep-sea hydrocarbon plume (Camilli et al., 2010; Hazen et al., 2010; Valentine et al., 2010,2012; Redmond and Valentine, 2011; Baelum et al., 2012; Mason et al., 2012), and in particular a rapid response of members of the Gammaproteobacteria to hydrocarbon inputs. Specifically, there was an initial increase in relative abundance of members of the Oceanospirillales (Hazen et al., 2010; Redmond and Valentine, 2011; Mason et al., 2012), followed by members of the generaColwellia and Cycloclasticus during later sampling periods (Redmond and Valentine, 2011;Valentine et al., 2012; Dubinsky et al., 2013).

Comparably less is known about the fate of the oil that reached the shore during the Deepwater Horizon spill. One study by Kostka et al. (2011) investigated the impact of the oil on beach samples collected several months after the spill occurred (July and September 2010) at municipal Pensacola Beach, Florida. By 16S rRNA gene sequencing, the authors found that the spill had a significant impact on the abundance and community composition of indigenous bacteria in beach sand with increases in many members of the Alpha- and Gammaproteobacteria, including some well-known hydrocarbon degraders (Alcanivorax and

Marinobacter) (Yakimov et al., 1998; Alonso-Gutiérrez et al., 2009). In the same study, several proteobacterial isolates, capable of growth on oil as their sole carbon source, were obtained from the contaminated samples (Kostka et al., 2011).

Here we aimed to determine the response of indigenous beach microbial communities to the oil that washed ashore early in the spill history. We focused our efforts on Elmers's Beach, Grand Isle, LA. This location was one of the most heavily oiled beaches in the Gulf, where oil began washing up onto the beach in early May 2010 (OSAT-2, 2011). A total of 153 oil contaminated and uncontaminated samples were collected at three time points in June 2010, while the oil continued to accumulate on the beach. The well was finally capped on July 15, 2010 and declared sealed on September 19, 2010.

We performed targeted 16S rRNA gene sequencing and total RNA sequencing (metatranscriptomics) to determine the composition of the microbial community, as well as to elucidate which members were actively degrading hydrocarbons in oiled samples. In addition, we isolated putative MC252 oil degrading microorganisms and studied their potential for hydrocarbon degradation. This study revealed a succession in the microbial community structure on the beach during early time points in the Deepwater Horizon oil spill. This study also represents the first use of metatranscriptome data to highlight the expression of genes involved in hydrocarbon transformations in a coastline community.

MATERIALS AND METHODS

Sample Collection

Beach sand cores were collected on Elmer's Beach (29.1782853, −90.0684072) at three time points on 03/06/2010 (n = 7), 21/06/2010 (n = 7), and 29/06/2010 (n = 3). Sand cores (10–20 cm deep) were taken by manual insertion of 40 cm long polybutyrate plastic liners into the sand. The cores were taken from locations submerged in the water close to the waterline, at the waterline, and inland. To circumvent potential contamination from the polybutyrate liners, each sand core was sub-cored using a 25 mm diameter sterile copper pipe, and sectioned into 3 cm depth intervals. Additionally, tar-like samples found on the surface of the beach (n = 24) were collected at each sampling period by aseptically scraping approximately 2–10 g into sterile 50 mL conical tubes. All samples were kept on ice in the field and were maintained at 4°C until further processing.

Acridine Orange Direct Counts

Approximately 1 g of each sample was homogenized and diluted in 1X PBS. Samples were filtered through a 0.2 µm pore size black polycarbonate membrane (Whatman International Ltd., Piscataway, NJ). Filtered cells were stained with 25 mg/mL acridine orange for 2 min in the dark. Unbound acridine orange was filtered through the membrane with 10 mL filter sterilized 1X PBS (Sigma Aldrich Corp., St. Louis, MI) and the rinsed membrane was mounted on a slide for microscopy. Cells were imaged with a FITC filter on a Zeiss Axioskop (Carl Zeiss, Inc., Germany).

Petroleum Hydrocarbon Concentrations

Total petroleum hydrocarbon (TPH) concentrations were determined using previously published procedures (Hazen et al., 2010) with the following modifications: 500 µL of chloroform were added to 500 mg of sample and then vortexed thoroughly, shaken for 2 min and sonicated for 2 min. The samples were incubated at room temperature for 1 h, centrifuged at 2,000 rpm for 5 min, and 50 µL of the extract was removed for analysis on an Agilent 6890N GC/FID (Santa Clara, CA). The GC was operated with an injector temperature of 250°C and detector temperature of 300°C, following a temperature program of 50°C for 2 min, ramped by 5°C/min until reaching 300°C and subsequently held for 15 min. TPH were quantified by integrating all the peaks from 20 to 60 min and comparing to oil standards (0–200 mg/L) obtained from the Macondo source well during the Deepwater Horizon spill.

PAH and alkane compound analysis was completed on the Agilent 6890N equipped with a 5972 mass selective detector and operated in SIM/SCAN mode. The injection temperature was 250°C, detector temperature was 300°C, and column used was 60 m Agilent HP-1 MS with a flow rate of 2 mL/min. The oven temperature program included a 50°C hold for 3 min ramped to 300°C at 4°C/min with a final 10 min hold at 300°C. Compound identification was determined from selective ion monitoring coupled with comparison to known standards and compound spectra in the NIST 08 MS library. Biomarker profiles were obtained by running the same samples in SIM mode targeting ions 191 for hopanes and 217 for steranes. Monitoring these ions has been widely used for oil source identification and degree of biodegradation (Venosa et al., 1997; Volkman et al., 1983; Greenwood and Georges, 1999; Hauser et al., 1999; Rosenbauer et al., 2010) and was utilized here to compare oil biomarker fingerprint to oil from the MC 252 source oil (Macondo crude). A proxy for biodegradation within the samples was calculated using the depletion of C_{25} with respect to C_{17} and the ratio of branched to aliphatic alkanes.

DNA EXTRACTION

Samples were extracted in duplicate using a modified Miller DNA extraction method (Miller et al., 1999). Approximately 0.5 g of each sample was placed into an FT500-ND PulseTube (Pressure BioSciences, Inc., USA). 300 µL of Miller phosphate buffer and 300 µL of Miller SDS lysis buffer were added and mixed. 600 µL phenol:chloroform:isoamyl alcohol (25:24:1) were then added, and the tubes were subjected to 25 cycles of 35,000 psi for 10 s and ambient pressure for 10 s, to improve cell lysis. The mixture was transferred to a Lysing Matrix E tube (MP Biomedicals, Solon, OH) and subjected to bead-beating at 5.5 m/s for 45 s in a FastPrep (MP Biomedicals, Solon, OH) instrument. The tubes were centrifuged at 16,000 × g for 5 min at 4°C and 540 µL of supernatant was transferred to a 2 mL tube with addition of an equal volume of chloroform. Tubes were mixed and then centrifuged at 10,000 × g for 5 min, after which 400 µL of the aqueous phase was transferred to another tube and 2 volumes of Solution S3 (MoBio, Carlsbad, CA) were added and mixed by inversion. The subsequent clean-up methods were based on the MoBio Soil DNA extraction kit according to the manufacturer›s instructions. Samples were recovered in 60 µL 10 mM Tris and stored at −20°C.

Community Profiling and Statistical Methods

Small subunit (SSU) rRNA gene sequences were amplified from duplicate DNA extractions using the primer pair 926f/1392r as previously described (Kunin et al., 2010). The reverse primer included a 5 bp barcode for multiplexing of samples during sequencing. Emulsion PCR and sequencing of the PCR amplicons was performed at DOE's Joint Genome Institute following manufacturer's instructions for the Roche 454 GS FLX Titanium technology, with the exception that the final dilution was 1e-8 (Allgaier et al., 2010).

Sequence reads were submitted to the PyroTagger computational pipeline (Kunin and Hugenholtz, 2010) where the reads were quality filtered, trimmed, dereplicated and clustered at 97% sequence identity. OTU tables generated from Pyrotagger were then imported into the QIIME pipeline (Caporaso et al., 2010) for further analyses. The number of sequence reads in each sample varied, therefore, the dataset was rarified. Alpha diversity calculations were performed on rarified data.

Multivariate community analysis was performed within PCORD 5 software (McCune et al., 2002) using normalized OTU data (family-level and OTU level). OTUs found in less than two samples were removed. Outliers were removed from the dataset using PCORD 5 with a cutoff of two standard deviations. The Bray-Curtis distance measure was used for non-metric multidimensional scaling (nMDS). Pearson correlation coefficients were calculated for metadata variables and each axis of the nMDS.

RNA Extraction, Amplification, and Sequencing

Total RNA was extracted from three of the oil contaminated samples as previously described (Kasai et al., 2001) and amplified using the Message Amp II-Bacteria Kit (Ambion, Austin, TX) following the manufacturer's instructions. First strand synthesis of cDNA from the resulting antisense RNA was carried out with the SuperScript III First Strand Synthesis System (Invitrogen, Carlsbad, CA). The SuperScript Double-Stranded cDNA Synthesis Kit (Invitrogen, Carlsbad, CA) was used to synthesize double stranded cDNA. cDNA was purified using a QIAquick PCR purification kit (Qiagen, Valencia, CA) and poly(A) tails were removed by digesting purified DNA with BpmI for 3 h at 37°C. Digested cDNA was purified with QIAquick PCR purification kit (Qiagen, Valencia, CA). RNA was prepared for sequencing using the Illumina Truseq kit following the manufacturer's guidelines. Each library was sequenced on one lane of the Illumina HiSeq platform using the 150 bp Paired-end technology resulting in a total of 57 Gb of sequence data for all three samples.

Metatranscriptomics Data Analysis

Raw Illumina sequence reads from each of three surface-contaminated samples (one from each sampling date) were trimmed using the CLC Genomics Workbench v5.0.1 with a quality score limit of 0.05. Phred quality scores (Q) were imported into the genomics workbench, where they were converted to error probabilities, using $p_{error} = 10^{Q/-10}$ and were trimmed using a limit of 0.05 as described in the CLC Workbench Manual (http://www.clcbio.com).

Sequences shorter than 50 bp in length and all adapter sequences were removed. To characterize the active microbial community members, unassembled reads were searched against the Greengenes (DeSantis et al., 2006) database of 16S rRNA genes using BLASTn with a bit score cutoff of >100.

Transcript profiles from each sample were determined by first subjecting trimmed unassembled reads from each sample to ORF calling using Prodigal (Hyatt et al., 2010). Resulting ORFs were compared to a translated in-house hydrocarbon gene database using BLASTp. This database was constructed using all KEGG genes involved in hydrocarbon degradation from the KEGG database (Kanehisa and Goto, 2000). For the resulting BLAST outputs, the highest bit score was selected (min bit score >40). Metatranscriptome data from each sample were normalized to RecA expression levels. A pairwise statistical comparison of the BLAST analyses was carried out using STAMP (Parks and Beiko, 2010) using a two-sided Chi-square test (with Yates correction) statistic with the DP: Asymptotic-CC confidence interval method and the Bonferroni multiple test correction. A p-value of < 0.05 was used with a double effect size filter (difference between proportions effect size < 1.00 and a ratio of proportions effect size < 2.00. The metatranscriptome from the June 29 sampling date yielded an insufficient number of transcripts after quality filtering, thus subsequent analyses of the metatranscriptome data focused on the June 3 and June 21 samples.

Paired-end Illumina reads from each of the June 3 and June 21 samples were assembled using theDe Novo Assembly Tool within the CLC Genomics Workbench at a word size of 20 and a bubble size of 50. Reads were scaffolded onto the contigs, which were submitted to MG-RAST (Meyer et al., 2008) for annotation. In MG-RAST, functional tables were generated for each sample against the KO annotation database, using default parameters (1e-5 maximum e-value cutoff, 60% minimum sequence identity, and 15 bp of minimum alignment length). To determine which organisms express genes involved in hydrocarbon degradation, contigs for each enzyme mapping to a xenobiotic pathway were annotated against the M5NR database for best-hit organismal classification using the default parameters. Using default parameters for the best-hit classification tool in MG-RAST, contigs were annotated against the Greengenes database to further assess presence of microbial community members through 16S rRNA transcripts. Recruitment plots were generated using a maximum e-value cutoff of 1e-3 and a log_2 abundance scale. Contigs mapping to xenobiotic pathways were rarefied to a depth of 20,000 annotated contigs each. Xenobiotic degradation maps annotated using Kegg Orthology (KO) were downloaded from the KEGG server in KGML format and manually colored using the KGML editor (Klukas and Schreiber, 2007). Charts were generated from the Krona template (Ondov et al., 2011).

Assembled data are publicly available in the MG-RAST database under project ID 7309. Raw reads were submitted to NCBI's sequence read archive under project ID SUB442498.

Enrichments and Isolations

Bacteria were isolated from sand cores and contaminated beach samples after incubation under aerobic conditions with 100 ppm Macondo oil (MC 252) in either Marine broth medium (Difco), Minimal marine medium (Baelum et al., 2012) or Synthetic minimal marine medium. Synthetic Minimal marine medium was prepared as follows: For 1 L, autoclaved separately 850 mL of 20 g NaCl, 0.67 g

KCl, 10 mL each of mineral and vitamin mixes (Coates et al., 1995), 100 mL of 30 mM phosphate buffer (pH 7.5), and added to 50 mL of 10.1 $MgCl_2.6H_2O$ + 1.52 g $CaCl_2$. 2 H_2O. Enrichments that resulted in an increase in turbidity, in addition to an increase in cell number by microscopic observations, were transferred periodically into fresh media. After 3–4 transfers, colonies were obtained by plating on the respective agar plates and were incubated for 1 week. Isolates were obtained from single colonies and incubated aerobically in modified Synthetic Seawater medium with 100 ppm MC252 oil as the sole carbon source. Within a few days, the oil initially observed as a thin layer floating on top disappeared with a concurrent increase in cell number. At this point, DNA was extracted from the cultures using the MoBio UltraClean Microbial DNA Isolation Kit (MoBio Inc, Carlsbad, CA). PCR amplification was conducted using universal bacterial 16S rRNA gene primers 27F and 1492R in 50 ul reactions, with a final concentration of 0.025 unit/µl Taq, 0.2 mM dNTPs, 15 ng of DNA template, and 0.04 µM primer. Initial denaturation was at 95°C for 180 s, followed by 25 cycles of melting at 95°C for 30 s, annealing at 54°C for 30 s, extension at 72°C for 60 s. This was followed by a final extension of 10 min at 72°C and samples were held at 4°C on completion of amplification. Verified 16S amplicons were purified using the procedure provided in the MoBio Ultraclean PCR Clean-up kit (MoBio, Carlsbad, CA). Samples were sequenced using the Applied Biosystems ABI 3730XL DNA Analyzers with the BigDye Terminator V3.1 Cycle Sequencing Kit (Applied Biosystems, Carlsbad, CA), according to the manufacturer›s instructions.

Oil Degradation with Isolates

Different selective minimal media were prepared to test individual isolates for their ability to degrade oil, since the isolates belonged to different genera and had different nutritional requirements. Marinobacter isolate 33 was grown in MC252 oil amended with minimal marine media. Roseobacter isolate 36 was grown in modified Sistrom's Minimal Medium (Sistrom, 1962). Oil degradation experiments were set up in 30 mL of respective media amended with 20 ppm MC252 oil and 0.1 ppm COREXIT 9500, inoculated with the respective bacterial cultures, and incubated at room temperature in the dark. The inoculant was grown in the respective minimal medium amended with 0.1%

Yeast extract to promote biomass. Prior to inoculation, cells were pelleted and washed in phosphate buffer (pH 7.5) to remove any carry over of media constituents. Heat killed cells (autoclaved) were used as negative controls, by 10% inoculation into experimental bottles containing oil and media.

At periodic intervals during the incubations, experimental bottles were sacrificed for hydrocarbon analyses to determine the extent of oil degradation. All glassware used in extraction and analyses was muffled at 500°C for 4 h prior to use. To extract hydrocarbons, the entire culture volume (30 ml) was transferred from the experimental bottles to a 50 mL glass culture tube with a Teflon-lined lid. The empty bottles were extracted three times with 2 mL of chloroform (BDH, ACS grade) to assay hydrocarbons sorbed to the glass and the rinses were added to the 50 mL tube. This mixture was vortexed for 1 min and extracted for 1 h, after which they were re-vortexed and centrifuged at 2000 rpm for 15 min to aid the separation of the chloroform from the aqueous media layer. The chloroform layer was removed with a glass pipette into a GC vial and analyzed as described above.

RESULTS

State of the Sampling Site

On June 3, 2010, the sampling site was almost completely covered to the tidal berm with viscous oil. Seawater washing up on the shore contained large, amorphous globules of oil. On June 21, 2010, the beach no longer contained visible globules of oil and the surface of the sampling site was no longer covered in oil. Instead, the oil present was in the form of small dried globules, less than 2 cm in diameter. By June 29, 2010, oil and oil mixed with foam were evident at the sampling site. The beach surface was rust in color and a light sheen of oil was noted on the seawater surface.

Chemical Analysis

The hydrocarbon profiles of the beached oil and contaminated sand core samples showed a clear correspondence to the MC252 oil (Supplemental Figure. S1). Total petroleum hydrocarbons (TPH) ranged from 0 mg/kg to 2072 mg/kg. Several components in the oil decreased over time and were significantly depleted by June 21 and June 29 sampling dates. Specifically, there was a depletion of shorter alkanes ($C_{17}-C_{20}$) and a corresponding higher relative amount of longer chain alkanes ($>C_{20}$) and branched alkanes. Cluster analysis of hydrocarbons revealed a clustering of the samples according to the level of hydrocarbon contamination (Supplemental Figure. S2). PAHs were detected in more than one third of the contaminated samples. Three-ring PAHs including, fluorene, anthracene, and phenanthracene and four-ring PAHs, including chrysene and pyrene, were highest in concentration of the measured PAH compounds, while naphthalene and other two-ring compounds were present in lower amounts and were nearly completely depleted in the less contaminated and uncontaminated samples from all time points.

Microbial Community Analyses

Cell counts ranged from 10^5 cells g^{-1} in uncontaminated samples to more than 10^7 cells g^{-1} in highly contaminated, beached oil samples and this difference was significant (t-test; $p = 2.97 \times 10^{-5}$). Therefore, there was a significant increase in microbial cell density as a result of the hydrocarbon influx on the beach, as previously reported by (Kostka et al., 2011).

We retrieved >1.6 million non-chimeric, quality filtered 16S rRNA gene sequences from a total of 153 oiled and uncontaminated samples, yielding more than 11,000 sequences per sample. The sequence data were dominated by OTUs corresponding to Alpha- and Gammaproteobacteria (Figure 1). Several OTUs that were abundant in the oil-contaminated samples corresponded to taxa with members known to degrade hydrocarbons, including Rhodobacteraceae, Alter omondaceae,Pseudomonadaceae, Chromatiaceae, Alcanivoraceae, and other families within theOceanospiralles. Samples with the highest concentrations of hydrocarbons had higher relative abundances of Alphaproteobacteria (Figure 1).

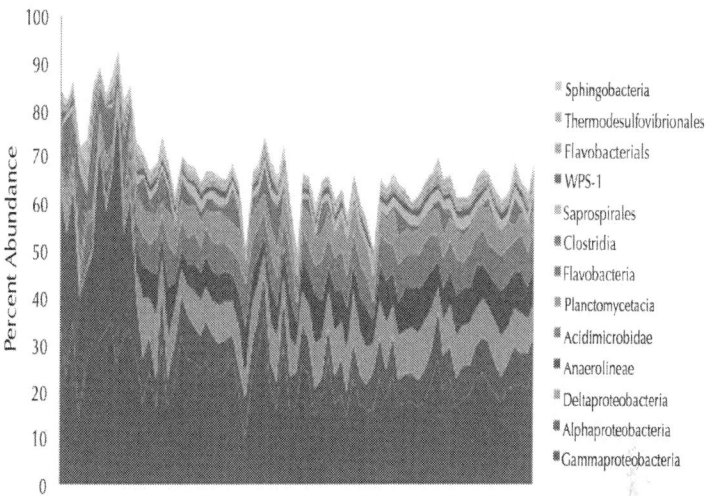

Figure 1: Percent abundance of the 13 most abundant bacterial classes using 16S rRNA gene sequences. Samples are ordered from highest to lowest TPH concentration, left to right.

Non-metric multidimensional scaling (nMDS) analysis revealed a pronounced response of the microbial community to oil contamination (Figure 2). Samples with high TPH concentrations clustered separately from less contaminated samples (Pearson correlation to Axis 1; r = 0.971). In addition, the TPH concentration was inversely related to several alpha diversity measures. Co-inertia analyses revealed that the microbial communities differed significantly between the two types of contaminated samples: beached oil and oil-contaminated sand (p-value < 0.001). The beached-oil samples also clustered separately by time (Pearson correlation to Axis 1; r = 0.869), suggesting temporal shifts in the microbial community as a response to the oil spill The depletion in TPH was also positively correlated with time of sampling for all of the contaminated samples. Shifts in the microbial community aligned with continuous disappearance of hydrocarbons during the sampling period. Several PAHs and aliphatic hydrocarbon components were among the highest factors that correlated to Axis 2 on the nMDS plots Pearson correlations revealed that Rhodobacteraceae and Alteromonadaceae were most highly correlated with hydrocarbon concentrations in the contaminated samples with genus and species-specific differences within sand and beached oil matrices. For example, sequences with closest homology to Rhodobacter sp., Jannachia sp., and Marinobacter lutaoensis had the highest correlation to beached oil samples), while Ruegeria sp., Jannachia sp., Alishewanella baltica, and Pseudomonas pachastrellae correlated with contaminated sand samples Of these highest correlating OTUs, the Marinobacter and Pseudomonasgenera were the most prevalent and abundant OTUs in the dataset, comprising up to 7 and 4% of the total community, respectively. It should also be noted, that microbial community composition and hydrocarbon profiles were highly correlated (Mantel test; t > 0, p = 0.00000, r = 0.6104).

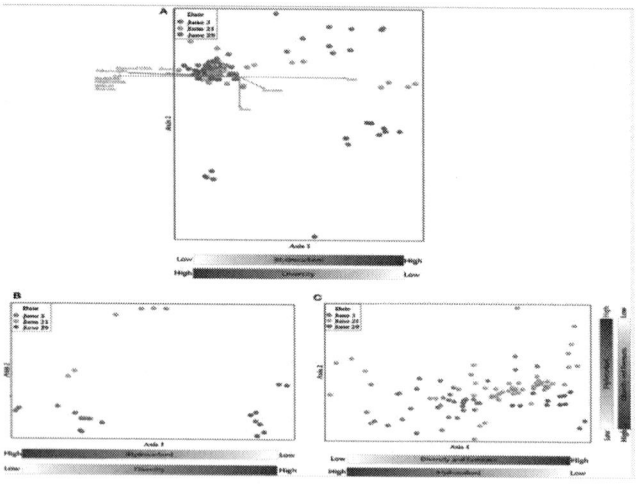

Figure 2: (A) Non-metric multidimensional scaling ordination of beached oil and sand samples based on the relative abundance pyrotag sequences assigned to family-level taxonomy. The ordination plot was rotated to maximize the degree of correlation with the total petroleum hydrocarbon variable. A two dimensional solution was found and the final stress was 0.023. (B) Non-metric multidimensional scaling ordinations of beached oil based on the relative abundance pyrotag sequences assigned to family-level taxonomy. The ordination plot was rotated to maximize the degree of correlation with the time variable. A two dimensional solution was found and the final stress was 0.039. (C) Non-metric multidimensional scaling ordinations of sand samples based on the relative abundance pyrotag sequences assigned to family-level taxonomy. The ordination plot was rotated to maximize the degree of correlation with the time variable. A two dimensional solution was found and the final stress was 0.086.

Metatranscriptomics of Oil Contaminated Samples

In order to assess which hydrocarbon degradation genes were expressed, we studied the metatranscriptomic profiles of representative heavily oiled samples. Approximately 380 million paired end sequences (57 Gb) were retrieved from three beached oil samples, one from each sampling date (June 3, June 21, June 29). Our goal was to determine what types of genes were expressed in the beach community as a whole in response to heavy oil contamination. We found that 40–67% of the quality filtered reads contained ribosomal RNA genes, which was not surprising considering rRNA depletion was not applied to these samples prior to sequencing, given the low RNA yields. When analyzing which taxa were most prevalent in the rRNA from the metatranscriptomes, we saw similar trends to the 16S rRNA microbial community analysis. Metatranscriptome data matching the Greengenes SSU database were dominated by the proteobacteria (74%), more specifically the Alteromonadales (30%), Oceanospirillales (11%), and the Rhodobacterales (8%). Further, we found that when the metatrascriptome data were compared to the SSU Greengenes database, 19.2% of sequences annotated at the genus level matched toMarinobacter.

Even though the samples were dominated by ribosomal genes, more than 100 million of the quality filtered reads were available for functional gene annotation. Nearly 17 million of these reads matched to the hydrocarbon gene database. A total of 3553 different matches to the hydrocarbon database were retrieved from the metatranscriptomics data with an average of 2357 reads mapping to each hit. Comparison of the unassembled data to the hydrocarbon gene database revealed that enzymes involved in degradation of a variety of hydrocarbons, including PAHs were expressed; including a variety of monoxygenases and dioxygenases, and those involved in converting PAHs to dihydrodiols Genes involved in the pathway for gentisate and substituted gentisate degradation were also expressed. Gentisate is a central metabolite in the aerobic biodegradation of both simple and complex aromatic hydrocarbons.

Two of the metatranscriptomes were assembled (those from the June 3 and June 21 sampling dates) yielding approximately 350,000 and 150,000 contigs (>150 bp), respectively, and the assemblies were also screened for hydrocarbon degradation genes. When the metatranscriptomes were searched for matches to reference genomes in the MG-RAST database, Marinobacter aquaeolei strain VT8 was the closest match (94% average identity) (Figure 3). The most abundant xenobiotic degradation transcripts and overall functional transcripts matching to this strain were cyclohexanone monooxygenase, naphthyl-2-methylsuccinyl-CoA dehydrogenase, naphthyl-2-methylsuccinyl-CoA dehydrogenase, 3-hydroxyacyl-CoA dehydrogenase/ enoyl-CoA hydratase, and a succinate dehydrogenase complex (Table 1). Genes involved in motility were amongst the most abundant features of all contigs mapping to M. aquaeolei and included the CheA signal transduction histidine kinase involved in chemotaxis signaling and a flagellar hook-associated 2 domain-containing protein (Table 1).

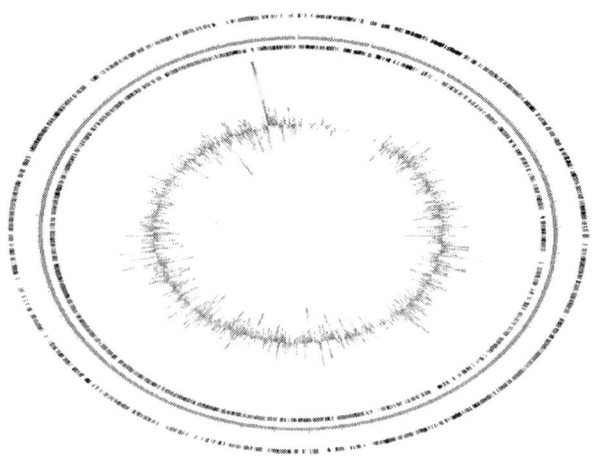

Figure 3: Recruitment of June 3 metatranscriptome to Marinobacter aquaeolei strain VT8, the organism to which the largest number of contigs mapped for both metatranscriptomes. The genome is approximately 4.8 Mb in size and the leading and lagging strands are represented by the outer most rings, separated by the blue ring, which indicates the position within the genome. Metatranscriptomic features are depicted as bar graphs inside the genome and their hit distribution is color-coded by e-value exponent as: blue, −3 to −5; green, −5 to −10; yellow, −10 to −20; orange, −20 to −30; red, less than −30. Figure was generated using the MG-RAST recruitment plot tool.

Table 1: Top xenobiotic and overall metatranscriptomic functions mapping to Marinobacter aquaeolei

Function	Relative bundance June 3*	Relative abundance June 21*
XENOBIOTIC		
Cyclohexanone monooxygenase	0.382	0.114
Naphthy1-2-methylsuccinyl-CoA dehydrogenase	0.318	0.795
Glutathione S-transferase	0.2.	0.000
3-hydroxyacyl-CoA dehydrogenase / enoyl-CoA hydratase	0.191	0.568
Succinate dehydrogenase	0.085	0.455
OVERALL		
CheA signal transduction histidine kinase	0.806	0.450
Flagellar hook-associated 2 domain-containing protein	0.467	0..0

| Elongation factor Tu | 0.2 | 1.023 |
| Tetratricopeptide TPR_4 | 0.361 | 0.909 |

* Relative abundances are percentage of total reads mapping to Marinobacter aquaelei.

Besides M. aquaeolei, xenobiotic degradation transcripts mapped to several other Proteobacteriaisolates in the MG-RAST database. For example, transcripts matched to PAH (Figure 4), n-alkane, and toluene degradation genes, matching to sequenced organisms in the Pseudomonadales, Burkholderiales, and Alteromonadales. Additionally, PAH (Figure 4), toluene, and benzoate pathways mapped to members of theRhodobacterales and toluene and benzoate metabolism transcripts mapped to Rhizobiales. It should be noted, that comparing metatranscriptomic data to KEGG pathways and organisms does not ascribe a complete pathway to a particular organism.

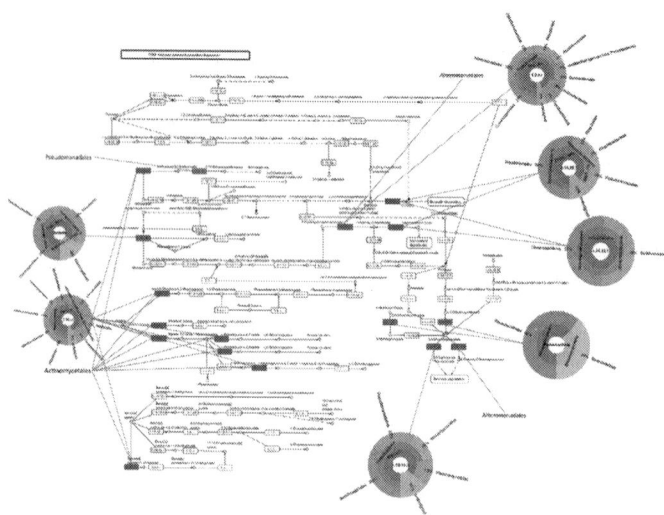

Figure 4: Polycyclic aromatic hydrocarbon degradation pathway. Assembled contigs are mapped to pathway from the KEGG database and colored in blue for June 3, red for June 21, and purple for presence in both time points. Pie charts indicating the best-hit taxonomic classification for each function were generated in Krona.

Isolation of Oil Degrading Strains from the Contaminated Beach Samples

Enrichment with oil-contaminated samples from the sampling location resulted in isolation of 18 unique bacterial strains belonging almost entirely to the Gammaproteobacteria. 16S rRNA gene sequencing revealed that almost half of the isolates shared highest sequence homology to members of the Pseudomonadales, including Pseudomonas stutzeri, Pseudomonas pachastrellae, andPseudomonas alcaligenes (Table 2). Three isolates belonging to the Marinobacter genus were retrieved from the more contaminated samples. Isolates having >99% sequence homology to knownAlcanivorax, Vibrio, Rheinheimera, and Bacillus sp. were also retrieved from these samples. Most of the isolates were halophilic, Gram-negative organisms, and showed the potential for degrading the MC252 oil.

Table 2: Cultured Isolates retrieved from beached oil and contaminated beach sands

Isolate number	Sample source	Phylogenetic order	Closest relative in greengenes 16S rRNA gene database (accession no)	Similarity (%)
	Sand	Vibrionales	*Vibrio* sp. str. C)Y102 MY174868.11	99.86
	Sand	Pseudomonadales	*Pseudomonas* sp. MOLA 58 1AM990833.11	99.64
	Sand	Vibrionales	*Vibrio* sp. str. CIY102 (AY174868.11	99.79
	Sand	Vibrionales	*Vibrio* sp. str. OY102 (AY174868.11	99.79
	Sand	Alteromondales	Marinobacter sp. str. Libra (AY734434.11 or *Marinobacter hydrocarbonoclasticus* str. JL795 (U512720.1)	99.93

12	Sand	Pseudomonadales	*Pseudomonas pachastrellae* str. PTG414 (EU603457.1)	97.14
14	Sand	Pseudomonadales	*Pseudomonas* sp. Da2 (AY570696.1)	99.43
16	Sand	Pseudomonadales	*Pseudomonas stutzeri* str. A1501 (NC_009434.1)	100
18	Sand	Bacillales	*Bacillus* sp. str. NRRL B-14911 (AA0X01000059.1)	99.72
19	Sand	Pseudomonadales	*Pseudomonas pseoloalooligenes* str. 14 (AB276371.1)	99.86
23	Sand	Chromatiales	*Rheinheimera* sp. 9712010) str. 97 (HM059656.11	99.64
25	Sand	Alteromondales	*Marinobacter* sp. str. Libra (AY734434.1) or *Marinobacter hydrocarbonoclasticus* str. JL795 (EF512720.1)	99.79
26	Sand	Pseudomonadales	*Pseudomonas pseudoalcaligenes* str. 14 (AB276371.1)	98.58
31	Beached oil	Oceanospirillales	*Alcanivorax* sp. str. Abu-1 (AB053129.1)	99.64
32	Beached oil	Pseudomonadales	*Pseudomonas* sp. str. BJCI-B3 (FJ600357.1)	94.52
33	Beached oil	Alteromondales	*Marinobacter* sp. Str. NT N31 (AB166980.1)	98.32
35	Beached oil	Rhodobacterales	*Citreicella thiooxidans* str. 2PR57-81EU440958.11	99.77
36	Beached oil	Rhoclobacterales	*Roseobacter* sp. str. 49Xb1 (EU090129.1)	99.93

Because of their high relative abundance in the 16S rRNA gene data in contaminated samples, two representative isolates, 33 (Marinobacter spp.) and 36 (Roseobacter spp.) were selected for their ability to grow using MC252 as the carbon source. Total hydrocarbons were extracted at selected time points and straight chain alkanes and PAHs (Figure 5) were depleted during the incubations for both cultures, although longer alkanes (C25 and longer) persisted after 15 and 20 days of incubation, respectively. It should be noted that the MC252 source oil used in the incubations was already depleted in the lighter hydrocarbons at the start of the incubation

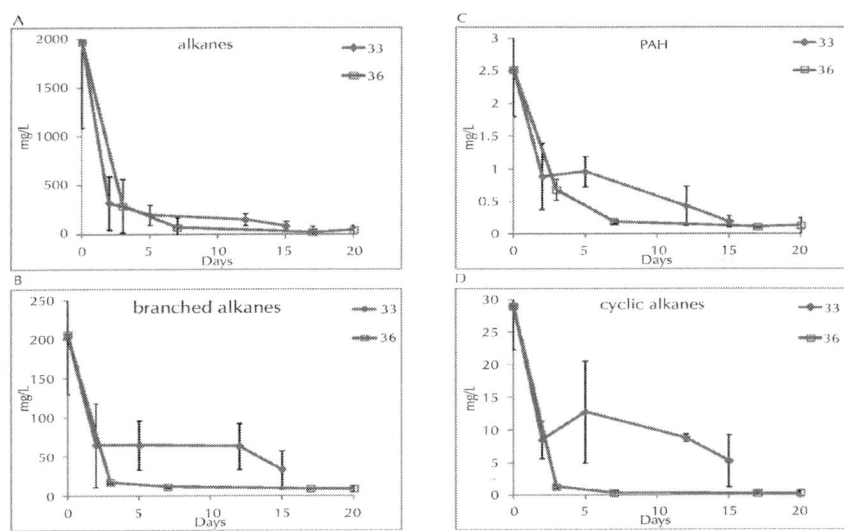

Figure 5: Loss of (A) straight alkanes, (B) branched alkanes, (C) PAHs, and (D) cyclic alkanes during incubation by isolate 33 (Marinobacter spp.) and isolate 36 (Roseobacter spp.).

DISCUSSION

Macondo oil from the Deepwater Horizon oil spill that reached the shore of Elmer's Beach caused shifts in the indigenous microbial communities in the beach sand toward a hydrocarbon-degrading consortium. This observation is consistent with previous studies that have assessed the impact of oil spills on coastline microbial communities (Kasai et al., 2001; Maruyama et al., 2003; Medina-Bellver et al., 2005; Alonso-Gutiérrez et al., 2009; Vila et al., 2010; Kostka et al., 2011). Kostka et al. (2011) also reported that highly contaminated samples exhibited higher bacterial cell densities than uncontaminated samples, and that there was a significant reduction in bacterial diversity associated with oil contamination. Here, we found that the contaminated samples collected from Elmer's Beach were generally dominated by Alpha- and Gammaproteobacteria (Figure 1) with up to 60% of the total microbial community being members of the Alphaproteobacteria. Other studies in the water column similarly reported a short-term shift of microbial communities toward specific members of the Gammaproteobacteria as an immediate response to crude oil inputs, which were then succeeded within 1 month by members of the Alphaproteobacteria (Abed et al., 2002; Röling et al., 2002;Hernandez-Raquet et al., 2006; Hazen et al., 2010; Redmond and Valentine, 2011; Valentine et al., 2012; Dubinsky et al., 2013).

Microbial community analysis revealed increases in the abundance of the Rhodobacteraceae andAlteromonadaceae in both the beached surface oil and contaminated beach sand samples. Therefore the different contaminated samples collected from the beach shared a similar bacterial community composition at the family level and exhibited parallel temporal successional changes in bacterial community structures driven by hydrocarbon inputs. During the first two sampling points, members of the Alteromonadaceae, with high sequence identity to Marinobacter lutaoensis, were very abundant in samples with high TPH concentrations. Members of the Marinobacter genus have previously been shown to be capable of degradation of both alkanes and PAH compounds with some isolates growing on single PAHs as their sole carbon source (Huu et al., 1999; Cohen, 2002; Shieh et al., 2003; Nicholson and Fathepure, 2004; Gerdes et al., 2005; Márquez and Ventosa, 2005; Brito et al., 2006; Gu et al., 2007; Cui et al., 2008; Rosano-Hernández and Fernández-Linares, 2009; Vila et al., 2010; Wu et al., 2010; Dos Santos et al., 2011). Here we also successfully isolated Marinobacterstrains from contaminated beach samples, which were capable of growth on MC252 oil as their sole carbon source. Several previous studies have reported the role of Marinobacter in degradation of oil (Gerdes et al., 2005; Vila et al., 2010; Kostka et al., 2011). The potential biodegradation of oil by these isolates at ambient temperature further supports their potential for natural biodegradation of oil in situ (Figure 5). However, it should be noted that further work is needed to determine the exact nature of the hydrocarbon transformations that occurred during the incubations and whether they were mineralized or transformed to other metabolites.

Several bacterial taxa within the Rhodobacteraceae were abundant in the highly contaminated samples. The Rhodobacteraceae are metabolically and ecologically diverse, comprised of photoheterotrophs that can grow, either, photoautotropically or chemotrophically, as well as chemoorganotrophs, fermenters, and methylotrophs. Several members of the Rhodobacteraceaehave previously been identified in oil polluted soils and marine environments and in fact have been shown to dominate oil polluted environments of the North Sea (Brakstad and Lødeng, 2005) and Southeast Asia (Harwati et al., 2008, 2009a,b). A few studies have demonstrated that the addition of photosynthetic bacteria to oil-polluted wastewater and soil triggers an increase in the abundance of hydrocarbon-oxidizing bacteria and thus enhances the rate of oil degradation (Martínez-Alonso et al., 2004; Llirós et al., 2008). Additionally, our cultivation-based experiments revealed that one representative of the Rhodobacteraceae, Roseobacter isolate 36, was also able to grow on MC252 as its sole carbon source. Overall, our data suggested oil degradation on the surface of beach sand that is exposed to light may have been promoted naturally by increases in photosynthetic populations.

Additionally many Pseudomonas species, having highest sequence homology to P. pachastrellae, were abundant in our 16S rRNA gene and cultivation experiments. Incidentally, similar pseudomonas strains were enriched from beach sands in the aftermath of both the Prestige oil spill in Northwestern Spain (Mulet et al., 2011) and other contaminated coastal sites during the Deepwater Horizon spill (Kostka et al., 2011) and these strains were shown to be central to the biodegradation of both aliphatic and aromatic hydrocarbons. Additionally, members of theAlcanivorax were abundant in the oil contaminated samples, corroborating previous 16S rRNA-based studies (Kasai et al., 2002; Kostka et al., 2011; Chakraborty et al., 2012).

Metatranscriptome analyses revealed that members of the Alpha- and Gammaproteobacteria were active in hydrocarbon degradation. This is the first study to determine functional genes involved in hydrocarbon degradation that were expressed in beach samples during the Deepwater Horizon spill. This study highlighted that metatranscriptomic data mapped to hydrocarbon degrading genes, including those involved in PAH, benzoate, and n-alkane degradation from Alteromonadales, Pseudomonales, and Rhodobacterales genomes. Data also mapped to other hydrocarbon degradation genes, including monooxygenases, dioxygenases, dehydrogenases, and hydratases, from members of these microbial classes. While this analysis doesn't necessarily ascribe a complete pathway to a particular organism, these results suggest that not only are these microorganisms abundant in the beach microbial community as suggested by the 16S rRNA gene data, but they may also play an active role in hydrocarbon degradation.

Marinobacter aquaeolei strain VT8 was the bacterium in the reference genome database that had the highest abundance of expressed genes in the oil contaminated samples, including those for cyclohexanone monooxygenase and naphthyl-2-methylsuccinyl-CoA dehydrogenase. In addition, transcripts for genes involved in chemotaxis and cellular motility mapped to Marinobactersuggesting that there was an active response to the hydrocarbon contamination in the beach communities, similar to the response observed for Oceanospirillales that were detected in the deep-sea plume (Mason et al., 2012). The high levels of gene expression observed for Marinobacter in the beach metatranscriptome data was supported by the finding that members of this genus were also enriched in the 16S rRNA data. In addition, we successfully isolated a representative ofMarinobacter from the contaminated beach samples and demonstrated the ability of the isolate to degrade MC252 oil. These data suggest that Marinobacter may have played a key role in degradation of the oil that reached the coast during the Deepwater Horizon oil spill.

CONCLUSIONS

During the Deepwater Horizon oil spill, MC252 oil originating from the Macondo well reached the coastline and Elmer's Beach was heavily impacted by the oil in June 2010, during which time we collected samples. Oil deposited on the shore appeared to cause a shift in the community structure toward a hydrocarbonoclastic consortia, as 16S rRNA gene sequencing demonstrated a diverse array of known petroleum hydrocarbon degrading microorganisms in these samples. Interestingly, several OTUs representative of previously described oil-degrading phototrophs were abundant in the heavily oiled samples from the first two sampling periods and these were succeeded by a diverse array of other potential oil-degrading bacteria. Metatranscriptome profiling revealed that members of theAlpha- and

Gammaproteobacteria expressed genes for hydrocarbon degradation in the contaminated samples, suggesting that they played a key role in potential degradation processes. Of note, Marinobacter were abundant members of the community in the oil-contaminated samples and expressed genes for degradation of hydrocarbons. Compared to other oil spills that have impacted shorelines, such as the Prestige oil spill that occurred in a cold pristine habitat, the disappearance of MC252 oil seemed more rapid. This difference in microbial response could be due to differences in temperatures between the two sites as well as differences in other environmental variables, including previous exposure to oil spills. Overall, this study of the microbial community response on the Gulf shoreline may assist in the understanding of microbial proxies for oil contamination in similar coastal ecosystems.

AUTHOR CONTRIBUTIONS

Regina Lamendella, Janet K. Jansson, and Terry C. Hazen were responsible for study conception and design. Regina Lamendella, Steven Strutt, and Janet K. Jansson were responsible for manuscript preparation. Sharon Borglin was responsible for chemical analyses. Romy Chakraborty was responsible for cultivation experiments. Regina Lamendella and Jenni Hultman were responsible for 16S RNA gene and metatranscriptomics experiments. Regina Lamendella, Steven Strutt, Olivia U. Mason, Emmanuel Prestat, and Neslihan Tas were responsible for bioinformatics and biostatistical analyses.

ACKNOWLEDGMENTS

This work was supported by a subcontract from the University of California at Berkeley, Energy Biosciences Institute to Lawrence Berkeley National Laboratory under its U.S. Department of Energy contract DE-AC02-05CH11231. The Energy Biosciences Institute to UC Berkeley is supported by a grant from British Petroleum. We acknowledge support from Theresa Pollard, Yvette Piceno and Dominique Joyner with ordering, and transportation of supplies and samples to and from the field.

REFERENCES

1. Abed, R. M. M., Safi, N. M. D., Köster, J., Beer, D. de, El-Nahhal, Y., Rullkötter, J. et al. (2002). Microbial diversity of a Heavily polluted microbial mat and its community changes following degradation of petroleum compounds. Appl. Environ. Microbiol. 68, 1674–1683. doi: 10.1128/AEM.68.4.1674-1683.2002
2. Allgaier, M., Reddy, A., Park, J. I., Ivanova, N., D'haeseleer, P., Lowry, S. et al. (2010). Targeted discovery of glycoside hydrolases from a switchgrass-adapted compost community. PLoS ONE 5:e8812. doi: 10.1371/journal.pone.0008812
3. Alonso-Gutiérrez, J., Figueras, A., Albaigés, J., Jiménez, N., Viñas, M., Solanas, A. M., et al. (2009). Bacterial communities from shoreline environments (Costa da Morte, Northwestern Spain) affected by the prestige oil spill. Appl. Environ. Microbiol. 75, 3407–3418. doi: 10.1128/AEM.01776-08
4. Baelum, J., Borglin, S., Chakraborty, R., Fortney, J. L., Lamendella, R., Mason, O. U., et al. (2012). Deep-sea bacteria enriched by oil and dispersant from the Deepwater Horizon spill. Environ. Microbiol. 14, 2405–2416. doi: 10.1111/j.1462-2920.2012.02780.x
5. Brakstad, O. G., and Lødeng, A. G. G. (2005). Microbial diversity during biodegradation of crude oil in seawater from the North Sea. Microb. Ecol. 49, 94–103. doi: 10.1007/s00248-003-0225-6
6. Brito, E. M. S., Guyoneaud, R., Goñi-Urriza, M., Ranchou-Peyruse, A., Verbaere, A., Crapez, M. A. C., et al. (2006). Characterization of hydrocarbonoclastic bacterial communities from mangrove sediments in Guanabara Bay, Brazil. Res. Microbiol. 157, 752–762. doi: 10.1016/j.resmic.2006.03.005
7. Camilli, R., Reddy, C. M., Yoerger, D. R., Van Mooy, B. A. S., Jakuba, M. V., Kinsey, J. C., et al. (2010). Tracking hydrocarbon plume transport and biodegradation at deepwater horizon. Science 330, 201–204. doi: 10.1126/science.1195223
8. Caporaso, J. G., Kuczynski, J., Stombaugh, J., Bittinger, K., Bushman, F. D., Costello, E. K., et al. (2010). QIIME allows analysis of high-throughput community sequencing data. Nat.

Methods 7, 335–336. doi: 10.1038/nmeth.f.303

9. Chakraborty, R., Borglin, S. E., Dubinsky, E. A., Andersen, G. L., and Hazen, T. C. (2012). Microbial Response to the MC-252 Oil and Corexit 9500 in the Gulf of Mexico. Front. Microbiol. 3:357. doi: 10.3389/fmicb.2012.00357

10. Coates, J. D., Lonergan, D. J., Philips, E. J. P., Jenter, H., and Lovley, D. R. (1995). Desulfuromonas palmitatis sp. nov., a marine dissimilatory Fe(III) reducer that can oxidize long-chain fatty acids. Arch. Microbiol. 164, 406–413. doi: 10.1007/BF02529738

11. Cohen, Y. (2002). Bioremediation of oil by marine microbial mats. Int. Microbiol. Off. J. Span. Soc. Microbiol. 5, 189–193. doi: 10.1007/s10123-002-0089-5

12. Cui, Z., Lai, Q., Dong, C., and Shao, Z. (2008). Biodiversity of polycyclic aromatic hydrocarbon-degrading bacteria from deep sea sediments of the Middle Atlantic Ridge. Environ. Microbiol. 10, 2138–2149. doi: 10.1111/j.1462-2920.2008.01637.x

13. DeSantis, T. Z., Hugenholtz, P., Larsen, N., Rojas, M., Brodie, E. L., Keller, K., et al. (2006). Greengenes, a Chimera-Checked 16S rRNA gene database and workbench compatible with ARB. Appl. Environ. Microbiol. 72, 5069–5072. doi: 10.1128/AEM.03006-05

14. Dos Santos, H. F., Cury, J. C., do Carmo, F. L., dos Santos, A. L., Tiedje, J., van Elsas, J. D., et al. (2011). Mangrove bacterial diversity and the impact of oil contamination revealed by pyrosequencing: bacterial proxies for oil pollution. PLoS ONE 6:e16943. doi: 10.1371/journal.pone.0016943

15. Dubinsky, E. A., Conrad, M. E., Chakraborty, R., Bill, M., Borglin, S. E., Hollibaugh, J. T., et al. (2013). Succession of hydrocarbon-degrading bacteria in the aftermath of the deepwater horizon oil spill in the Gulf of Mexico. Environ. Sci. Technol. 47, 10860–10867. doi: 10.1021/es401676y

16. Gerdes, B., Brinkmeyer, R., Dieckmann, G., and Helmke, E. (2005). Influence of crude oil on changes of bacterial communities in Arctic sea-ice. FEMS Microbiol. Ecol. 53, 129–139. doi: 10.1016/j.femsec.2004.11.010

17. Greenwood, P. F., and Georges, S. C. (1999). Mass spectral

characteristics of C19 and C20 tricyclic terpanes detected in Latrobe tasmanite oil shale. Eur. Mass Spectrom. 5, 221–230. doi: 10.1255/ejms.278

18. Gu, J., Cai, H., Yu, S.-L., Qu, R., Yin, B., Guo, Y.-F., et al. (2007). Marinobacter gudaonensis sp. nov., isolated from an oil-polluted saline soil in a Chinese oilfield. Int. J. Syst. Evol. Microbiol. 57, 250–254. doi: 10.1099/ijs.0.64522-0

19. Harwati, T. U., Kasai, Y., Kodama, Y., Susilaningsih, D., and Watanabe, K. (2008). Tranquillimonas alkanivorans gen.nov., sp. nov., an alkane-degrading bacterium isolated from Semarang Port in Indonesia. Int. J. Syst. Evol. Microbiol. 58, 2118–2121. doi: 10.1099/ijs.0.65817-0

20. Harwati, T. U., Kasai, Y., Kodama, Y., Susilaningsih, D., and Watanabe, K. (2009a). Tropicibacter naphthalenivorans gen. nov., sp. nov., a polycyclic aromatic hydrocarbon-degrading bacterium isolated from Semarang Port in Indonesia. Int. J. Syst. Evol. Microbiol. 59, 392–396. doi: 10.1099/ijs.0.65821-0

21. Harwati, T. U., Kasai, Y., Kodama, Y., Susilaningsih, D., and Watanabe, K. (2009b). Tropicimonas isoalkanivorans gen.nov., sp. nov., a branched-alkane-degrading bacterium isolated from Semarang Port in Indonesia. Int. J. Syst. Evol. Microbiol. 59, 388–391. doi: 10.1099/ijs.0.65822-0

22. Hauser, A., Dashti, H., and Khan, Z. H. (1999). Identification of biomarker compounds in selected Kuwait crude oils. Fuel78, 1483–1488. doi: 10.1016/S0016-2361(99)00075-7

23. Hazen, T. C., Dubinsky, E. A., DeSantis, T. Z., Andersen, G. L., Piceno, Y. M., Singh, N., et al. (2010). Deep-sea oil plume enriches indigenous oil-degrading bacteria. Science 330, 204–208. doi: 10.1126/science.1195979

24. Hernandez-Raquet, G., Budzinski, H., Caumette, P., Dabert, P., Le Ménach, K., Muyzer, G., et al. (2006). Molecular diversity studies of bacterial communities of oil polluted microbial mats from the Etang de Berre (France). FEMS Microbiol. Ecol. 58, 550–562. doi: 10.1111/j.1574-6941.2006.00187.x

25. Huu, N. B., Denner, E. B., Ha, D. T., Wanner, G., and Stan-Lotter, H. (1999). Marinobacter aquaeolei sp. nov., a halophilic bacterium isolated from a Vietnamese oil-producing well. Int. J. Syst. Bacteriol. 49(Pt 2), 367–375.

26. Hyatt, D., Chen, G.-L., LoCascio, P. F., Land, M. L., Larimer, F. W., and Hauser, L. J. (2010). Prodigal: prokaryotic gene recognition and translation initiation site identification. BMC Bioinformatics 11:119. doi: 10.1186/1471-2105-11-119
27. Kanehisa, M., and Goto, S. (2000). KEGG: kyoto encyclopedia of genes and genomes. Nucleic Acids Res. 28, 27–30. doi: 10.1093/nar/28.1.27
28. Kasai, Y., Kishira, H., Sasaki, T., Syutsubo, K., Watanabe, K., and Harayama, S. (2002). Predominant growth of Alcanivorax strains in oil-contaminated and nutrient-supplemented sea water. Environ. Microbiol. 4, 141–147. doi: 10.1046/j.1462-2920.2002.00275.x
29. Kasai, Y., Kishira, H., Syutsubo, K., and Harayama, S. (2001). Molecular detection of marine bacterial populations on beaches contaminated by the Nakhodka tanker oil-spill accident. Environ. Microbiol. 3, 246–255. doi: 10.1046/j.1462-2920.2001.00185.x
30. Kimes, N. E., Callaghan, A. V., Aktas, D. F., Smith, W. L., Sunner, J., Golding, B., et al. (2013). Metagenomic analysis and metabolite profiling of deep-sea sediments from the Gulf of Mexico following the Deepwater Horizon oil spill. Front. Microbiol. 4:50. doi: 10.3389/fmicb.2013.00050
31. Klukas, C., and Schreiber, F. (2007). Dynamic exploration and editing of KEGG pathway diagrams. Bioinformatics 23, 344–350. doi: 10.1093/bioinformatics/btl611
32. Kostka, J. E., Prakash, O., Overholt, W. A., Green, S. J., Freyer, G., Canion, A., et al. (2011). Hydrocarbon-degrading bacteria and the bacterial community response in Gulf of Mexico beach sands impacted by the deepwater horizon oil spill. Appl. Environ. Microbiol. 77, 7962–7974. doi: 10.1128/AEM.05402-11
33. Kunin, V., and Hugenholtz, P. (2010). PyroTagger: a fast, accurate pipeline for analysis of rRNA amplicon pyrosequence data. Open J. 1, 1–8. Available online at: http://www.theopenjournal.org/toj_articles/1
34. Kunin, V., Engelbrektson, A., Ochman, H., and Hugenholtz, P. (2010). Wrinkles in the rare biosphere: pyrosequencing errors can lead to artificial inflation of diversity estimates. Environ. Microbiol. 12, 118–123. doi: 10.1111/j.1462-2920.2009.02051.x
35. Llirós, M., Gaju, N., de Oteyza, T. G., Grimalt, J. O., Esteve, I.,

and Martínez-Alonso, M. (2008). Microcosm experiments of oil degradation by microbial mats. II. The changes in microbial species. Sci. Total Environ. 393, 39–49. doi: 10.1016/j.scitotenv.2007.11.034

36. Lubchenco, J., McNutt, M., Lehr, B., Sogge, M., Miller, M., Hammond, S., et al. (2010). Deepwater Horizon/BP Oil Budget: What Happened to the Oil? Available online at:http://www.noaanews.noaa.gov/stories2010/PDFs/OilBudget_description_%2083final.pdf

37. Márquez, M. C., and Ventosa, A. (2005). Marinobacter hydrocarbonoclasticus Gauthier et al. 1992 and Marinobacter aquaeolei Nguyen et al. 1999 are heterotypic synonyms. Int. J. Syst. Evol. Microbiol. 55, 1349–1351. doi: 10.1099/ijs.0.63591-0

38. Martínez-Alonso, M., de Oteyza, T. G., Llirós, M., Munill, X., Muyzer, G., Esteve, I., et al. (2004). Diversity shifts and crude oil transformation in polluted microbial mat microcosms. Ophelia 58, 205–216. doi: 10.1080/00785236.2004.10410228

39. Maruyama, A., Ishiwata, H., Kitamura, K., Sunamura, M., Fujita, T., Matsuo, M., et al. (2003). Dynamics of microbial populations and strong selection for Cycloclasticus pugetii following the Nakhodka oil spill. Microb. Ecol. 46, 442–453. doi: 10.1007/s00248-002-3010-z

40. Mason, O. U., Hazen, T. C., Borglin, S., Chain, P. S. G., Dubinsky, E. A., Fortney, J. L., et al. (2012). Metagenome, metatranscriptome and single-cell sequencing reveal microbial response to Deepwater Horizon oil spill. ISME J. 6, 1715–1727. doi: 10.1038/ismej.2012.59

41. Mason, O. U., Scott, N. M., Gonzalez, A., Robbins-Pianka, A., Bælum, J., Kimbrel, J., et al. (2014). Metagenomics reveals sediment microbial community response to Deepwater Horizon oil spill. ISME J. doi: 10.1038/ismej.2013.254. [Epub ahead of print].

42. McCrea-Strub, A., Kleisner, K., Sumaila, U. R., Swartz, W., Watson, R., Zeller, D., et al. (2011). Potential impact of the Deepwater Horizon oil spill on commercial fisheries in the Gulf of Mexico. Fisheries 36, 332–336. doi: 10.1080/03632415.2011.589334

43. McCune, B., Grace, J. B., and Urban, D. L. (2002). Analysis of

Ecological Communities. Corvallis, OR: MjM Software Design.
44. Medina-Bellver, J. I., Marín, P., Delgado, A., Rodríguez-Sánchez, A., Reyes, E., Ramos, J. L., et al. (2005). Evidence for in situ crude oil biodegradation after the prestige oil spill. Environ. Microbiol. 7, 773–779. doi: 10.1111/j.1462-2920.2005.00742.x
45. Meyer, F., Paarmann, D., D'Souza, M., Olson, R., Glass, E. M., Kubal, M., et al. (2008). The metagenomics RAST server – a public resource for the automatic phylogenetic and functional analysis of metagenomes. BMC Bioinformatics 9:386. doi: 10.1186/1471-2105-9-386
46. Miller, D. N., Bryant, J. E., Madsen, E. L., and Ghiorse, W. C. (1999). Evaluation and optimization of DNA Extraction and purification procedures for soil and sediment samples. Appl. Environ. Microbiol. 65, 4715–4724.
47. Mulet, M., David, Z., Nogales, B., Bosch, R., Lalucat, J., and García-Valdés, E. (2011). Pseudomonas diversity in crude-oil-contaminated intertidal sand samples obtained after the prestige oil spill. Appl. Environ. Microbiol. 77, 1076–1085. doi: 10.1128/AEM.01741-10
48. Nicholson, C. A., and Fathepure, B. Z. (2004). Biodegradation of benzene by halophilic and halotolerant bacteria under aerobic conditions. Appl. Environ. Microbiol. 70, 1222–1225. doi: 10.1128/AEM.70.2.1222-1225.2004
49. Ondov, B. D., Bergman, N. H., and Phillippy, A. M. (2011). Interactive metagenomic visualization in a Web browser. BMC Bioinformatics 12:385. doi: 10.1186/1471-2105-12-385
50. OSAT-2. (2011). Summary Report for Fate and Effects of Remnant Oil in the Beach Environment. Operational Science Advisory Team, Deepwater Horizon MC252. Prepared for Federal On-Scene Coordinator. Available online at:http://www.dep.state.fl.us/deepwaterhorizon/files2/osat_2_report__10feb.pdf
51. Parks, D. H., and Beiko, R. G. (2010). Identifying biologically relevant differences between metagenomic communities. Bioinforma. Oxf. Engl. 26, 715–721. doi: 10.1093/bioinformatics/btq041.
52. Redmond, M. C., and Valentine, D. L. (2011). Natural gas and temperature structured a microbial community response to the

Deepwater Horizon oil spill. Proc. Natl. Acad. Sci. U.S.A. doi: 10.1073/pnas.110875610

53. Röling, W. F. M., Milner, M. G., Jones, D. M., Lee, K., Daniel, F., Swannell, R. J. P., et al. (2002). Robust hydrocarbon degradation and dynamics of bacterial communities during nutrient-enhanced oil spill bioremediation. Appl. Environ. Microbiol. 68, 5537–5548. doi: 10.1128/AEM.68.11.5537-5548.2002

54. Rosano-Hernández, M. C., and Fernández-Linares, L. C. (2009). Bacterial diversity of marine seeps in the southeastern Gulf of Mexico. Pak. J. Biol. Sci. 12, 683–689. doi: 10.3923/pjbs.2009.683.689

55. Rosenbauer, R. J., Campbell, P. L., Lam, A., Lorenson, T. D., Hostettler, F. D., Thomas, B., et al. (2010). Reconnaissance of Macondo-1 Well Oil in Sediment and Tarballs from the Northern Gulf of Mexico shoreline, Texas to Florida. United States Geological Survey. Available online at: http://pubs.usgs.gov/of/2010/1290/

56. Shieh, W. Y., Jean, W. D., Lin, Y.-T., and Tseng, M. (2003). Marinobacter lutaoensis sp. nov., a thermotolerant marine bacterium isolated from a coastal hot spring in Lutao, Taiwan. Can. J. Microbiol. 49, 244–252. doi: 10.1139/w03-032.

57. Sistrom, W. R. (1962). The Kinetics of the Synthesis of Photopigments in Rhodopseudomonas spheroides. J. Gen. Microbiol. 28, 607–616. doi: 10.1099/00221287-28-4-607

58. Sumaila, U. R., Cisneros-Montemayor, A. M., Dyck, A., Huang, L., Cheung, W., Jacquet, J., et al. (2012). Impact of the Deepwater Horizon well blowout on the economics of US Gulf fisheries. Can. J. Fish. Aquat. Sci. 69, 499–510. doi: 10.1139/f2011-171.

59. The Federal Intragency Solutions Group: Oil Budget Calculator Science and Engineering Team (2010). Oil Budget Calculator Deepwater Horizon.

60. US Coast Guard, USGS, and NOAA. (2010). Deepwater Horizon MC252 Gulf Incident Oil Budget. Available online at:http://www.usgs.gov/foia/budget/08-02-2010.Deepwater%20Horizon%20Oil%20Budget.pdf

61. Valentine, D. L., Kessler, J. D., Redmond, M. C., Mendes, S. D., Heintz, M. B., Farwell, C., et al. (2010). Propane respiration

jump-starts microbial response to a deep oil spill. Science 330, 208–211. doi: 10.1126/science.1196830

62. Valentine, D. L., Mezic, I., Macesic, S., Crnjaric-Zic, N., Ivic, S., Hogan, P. J., et al. (2012). Dynamic autoinoculation and the microbial ecology of a deep water hydrocarbon irruption. Proc. Natl. Acad. Sci. U.S.A. 109, 20286–20291. doi: 10.1073/pnas.1108820109

63. Venosa, A. D., Suidan, M. T., King, D., and Wrenn, B. A. (1997). Use of hopane as a conservative biomarker for monitoring the bioremediation effectiveness of crude oil contaminating a sandy beach. J. Ind. Microbiol. Biotechnol. 18, 131–139. doi: 10.1038/sj.jim.2900304

64. Vila, J., María Nieto, J., Mertens, J., Springael, D., and Grifoll, M. (2010). Microbial community structure of a heavy fuel oil-degrading marine consortium: linking microbial dynamics with polycyclic aromatic hydrocarbon utilization. FEMS Microbiol. Ecol. 73, 349–362. doi: 10.1111/j.1574-6941.2010.00902.x

65. Volkman, J. K., Alexander, R., Kagi, R. I., and Rullkötter, J. (1983). GC-MS characterisation of C27 and C28 triterpanes in sediments and petroleum. Geochim. Cosmochim. Acta 47, 1033–1040. doi: 10.1016/0016-7037(83)90233-8

66. Wu, C., Wang, X., and Shao, Z. (2010). Diversity of oil-degrading bacteria isolated form the Indian Ocean sea surface. Wei Sheng Wu Xue Bao 50, 1218–1225.

67. Yakimov, M. M., Golyshin, P. N., Lang, S., Moore, E. R., Abraham, W. R., Lünsdorf, H., et al. (1998). Alcanivorax borkumensis gen. nov., sp. nov., a new, hydrocarbon-degrading and surfactant-producing marine bacterium. Int. J. Syst. Bacteriol. 48(Pt 2), 339–348. doi: 10.1099/00207713-48-2-339

68. Zukunft, P. (2010). Summary Report for Sub-Sea and Sub-Surface Oil and Disperion Detection: Sampling and Monitoring. Available online at: http://www.restorethegulf.gov/sites/default/files/documents/pdf/OSAT_Report_FINAL_17DEC.pdf

Chapter 4

Multiscale Erosion Surfaces of the Organic-Rich Barnett Shale, Fort Worth Basin, USA

Mohamed O. Abouelresh

Faculty of Petroleum and Mining Engineering, Suez University, Salah Naesem St., Etaka, Suez 43721, Egypt

ABSTRACT

The high frequency and diversity of erosion surfaces throughout the Barnett Shale give a unique view into the short-duration stratigraphic intervals that were previously much more difficult to detect in such fine-grained rocks. The erosion surfaces in Barnett Shale exhibit variable relief (5.08–61 mm) which commonly consists of shelly laminae, shale rip-up clasts, reworked mud intraclasts, phosphatic pellets, and/or diagenetic minerals (dolomite and pyrite) mostly with clay-rich mudstone groundmass. Several factors control this lithological variation, including the energy conditions, rate of relative sea-level fluctuation, rate of sedimentation, sediment influx, and the lithofacies type of the underlying as well as the overlying beds. The erosional features and their associated surfaces make them serve at least in part as boundaries between different genetic types of deposits but with different scales according to their dependence on base level and/or sediment supply. Accordingly, the studied erosion surfaces of the Barnett Shale can be grouped into three different scales of sequence stratigraphic surfaces: sequence-scale surfaces, parasequence-scale surfaces, and within trend-scale surfaces.

INTRODUCTION

The classic interpretation of organic-rich shale (≥0.5% total organic carbon) deposition emphasized continuous hemipelagic deposition in deep, quiet, low energy and stagnant basins, often with a stratified water column. However, Schieber [1] identified laterally continuous erosion surfaces in the Chattanooga Shale and he interpreted them as being the result of wave reworking and erosion of the sea floor. These surfaces are direct indications of major environmental events that may include nondeposition and/or erosion events.

The multiple erosion events, resultant surfaces, and erosion features are common within the coarser clastic rocks and are pivotal in classifying depositional sequences based upon relative sea-level fluctuations. On the other hand, in fine-grained sediments, these events are subtle [2], although the identification is important in elucidating and development of the sequence stratigraphic framework for such rocks.

Mud floored erosion in the geologic record is often associated with zones of intensely burrowed sediment; hiatus-concretions and prefossilized organic remains reveal complex cycles of exhumation and reburial associated with erosion (see [3–6]). Of particular significance are intervals of vertically mixed sediment, shells, and nodules associated with this erosion and/or reworking surfaces. Such sediment intervals yield a complete spectrum from in situ to extensively reworked shale clasts, thus allowing detailed reconstruction of the erosion event as an ongoing process; however, not all erosion surfaces are marked by shale clasts [7].

Noteworthy, the erosion surfaces and their resultant lags can provide an indirect measure of processes and palaeogeography that existed immediately preceding and during transgression, a time most commonly characterized by hiatus and erosional vacuity [8]. The erosion surfaces are relatively abundant throughout the Barnett Shale, punctuate the stacking patterns, and may obliterate the record of various strata [9]. However, erosion provides regional bounding surfaces for classifying this ~25 myr shale.

This work aims to identify the characters of erosion surfaces and their related features in organic-rich shales which occur on multiple scales from different erosional events. Moreover, this work also shows how these surfaces could be used as multiscale stratigraphic surfaces for classifying genetically related fine-grained rocks.

GEOLOGIC SETTING

The organic-rich Barnett Shale of Fort Worth Basin, Texas, is a world-class unconventional gas reservoir deposited over a period of 25 million years from 345 to 320 myr [10]. The Fort Worth Basin extends some 322–402 km from the Red River arch in the north to the Llano uplift in the south (Figure 1(a)). The Bend arch forms the western margin of the basin and the Ouachita thrust front forms the eastern margin. The basin covers approximately 24140 km^2. It is one of a series of foreland basins formed along the southern margin of the North American craton during the late Paleozoic Ouachita orogeny [11–13]. Paleozoic strata comprise almost the entire fill of the Fort Worth Basin. The basin fill thickens to the northeast, into the Oklahoma aulacogen, and thins to the south and west, toward the Llano uplift and the Bend arch, respectively. The Barnett Shale also thins to the west over the Bend arch where it interfingers with the Chappel Formation [14–17].

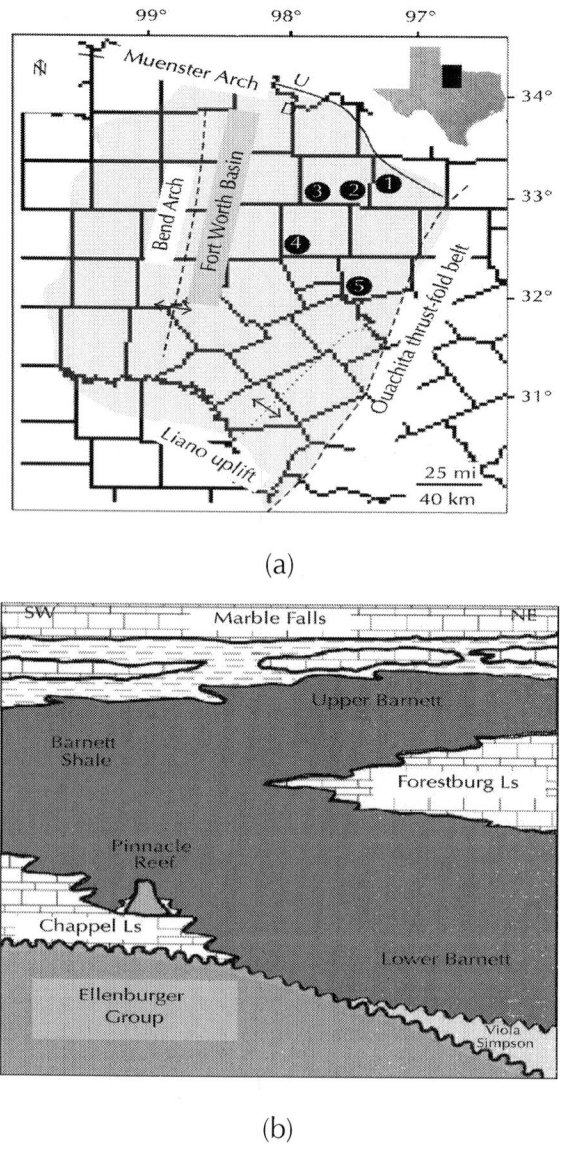

Figure 1: (a) Location map for the Fort Worth Basin showing the distribution of the Barnett Shale (shaded area) and structural and tectonic features; dots refer to the studied wells; 1: John Porter #3, 2: Sol Carpenter #7, 3: Adams SW #7, 4: Sugar Tree #3, and 5: Spencer Trussell 1-H (modified from [10]). (b) The schematic section shows an interpretation of the Mississippian stratigraphy (modified from [10]).

Following an extensive early Paleozoic transgression, erosion removed all of the Silurian and Devonian strata from the Fort Worth Basin [14]. The Barnett Shale was the first unit to be deposited when the seas returned (Figure 1(b)). The Barnett Shale was deposited on the karsted surface of the Ellenburger Group over a wide area [16]. Older stratigraphic studies suggest that most of the Barnett Shale accumulated either on a normal marine shelf [14] or in a relatively shallow, starved basin under euxinic conditions [16]. However, more recent work interprets the Barnett Shale in the Fort Worth Basin as deep-water slope-to-basin deposit [18, 19]. The relatively low energy and deep water environment, distant from a terrigenous sources, is inferred for the lower Barnett Shale. Whereas, depositional environment for the upper Barnett Shale is inferred to be shallower, proximal, well-oxygenated compared to the lower facies [9, 18].

Interestingly, the organic-rich Barnett Shale is characterized by a prolonged history of deposition and erosion due to frequent changes in relative sea level especially in the upper part; these fluctuations match with those proposed by Haq and Schutter [20] for the Late Mississippian. This type of fluctuation generated a variety of depositional, nondepositional, and erosional features, which have been placed within a sequence stratigraphic framework [9, 21].

DATA AND METHODOLOGY

The small grain size (<62 μm), subtle textural variations, lack of differences in erosion styles, and the gray to black color mean that the organic-rich Barnett Shale is difficult to characterize using conventional sedimentological techniques [22]. In this work, five continuous cores from the Barnett Shale (John Porter #3 (JP) in Denton County, Sol Carpenter H#7 (SC), Adams SW#7 (ASW) in Wise County, Sugar Tree #1 (ST) in Parker County and Spencer Trussell #1-H in Johnson County, Texas, USA) were studied.

Thirty three (33) thin sections collected from these cores represent a wide range of different erosion surfaces throughout the Barnett Shale. These thin sections were scanned using an Epson, 48-bit color, 4800 dpi scanner to give a higher resolution image than what can be obtained either from hand specimens or even under polarizing microscope. Core samples and thin sections observations were recorded and photographed in detail to shed light on the different characteristics of the erosional surfaces.

RESULTS

The detailed investigation of erosion surfaces throughout the Barnett Shale showed that these surfaces are part of zones vertical reliefs ranging from a few millimeters to several centimeters (Figure 2) as a result of differential erosion; considering that prior to compaction, this relief would have been up to 90% greater [23]. These zones consist of vertically mixed components within clay-rich mudstone groundmass. The most common components of the erosion surfaces include surface nature, shelly laminae, shale rip-up clasts, reworked intraclasts, phosphatic pellets, and diagenetic minerals (dolomite and pyrite). The following is a detailed description for each of these features which are also summarized in Table 1.

Table 1: Summary of the erosional features associated with the erosion surfaces

Relief	Ranges between 5 mm and 60 mm with an average 19 mm.
Lithology	Siliceous noncalcareous to calcareous mudstone groundmass, silty laminated, and phosphatic-rich mudstone. Pyrite is common and rarely dolomitized.
Surface nature Lower	Sharp to irregular, scoured and down-cutting contact.
Upper	Sharp, gradational to irregular, rarely disturbed with burrowing.

Matrix components	
Shelly laminae	Very thin (1–15 mm), singular to multiple, fragmented shelly laminae. Highly compacted (grain to grain contact), horizontally alignment. The most common types of shell fragments are bivalves, bryozoans, brachiopods, filling branch mollusks, and echinoderms with mainly calcite filling.
Shale rip-up clasts	Mainly flakey-like shapes with straight outlines. Others are rounded to subrounded, with the same lithology of the underlying facies. Occurs as suspended clasts in the overlying facies.
Reworked concretions	In situ sandy-size to transported gravely-size concretions. Rounded, subrounded to irregular clasts. Slightly to highly compacted and subhorizontally oriented. Shaley to silty and phosphatic internal lithology. Pyritic effect is very common.
Phosphatic pellets	Very common, rounded to subrounded with multiple internal cores, sometimes broken, irregular, rarely elongated with their long axis parallel to bedding planes. They range in size from less than 1 mm up to 1 cm. The nuclei of the phosphatic intraclasts may include shell fragments, detrital quartz, and glauconite grains.
Pyrite	Common as framboidal shape and rarely as fine euhedral crystal. Not in all samples.

88 Deepwater Petroleum Exploration & Production

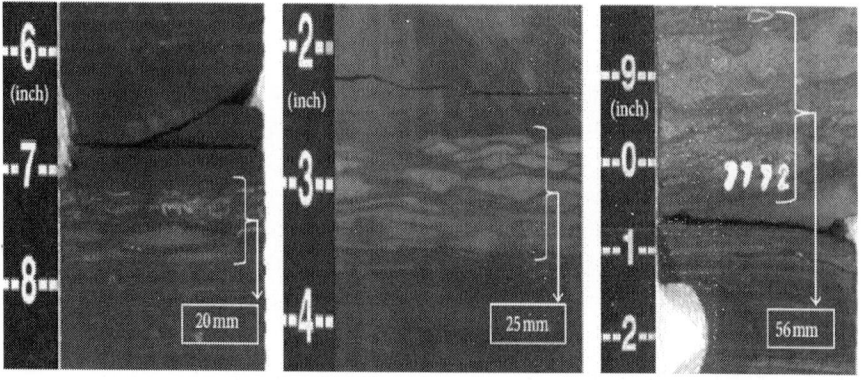

Figure 2: Core photos show different reliefs of erosion surfaces and their related features ranging from 20 mm to 56 mm.

Surface Nature

Description

The nature of erosion surfaces is frequent and identifiable on thin sections and core samples. The detailed investigation of the studied cores showed that these surfaces exhibit different shapes including (Table 1) sharp straight (though sharp irregular has been seen) (Figure 3(a)), scour, irregular (Figure 3(b)), and gradational upper contact (Figure 3(c)). In some cases the surface may down-cut into older deposits (Figure 3(d)). The lack of burrowing with most of lower sharp contacts is common.

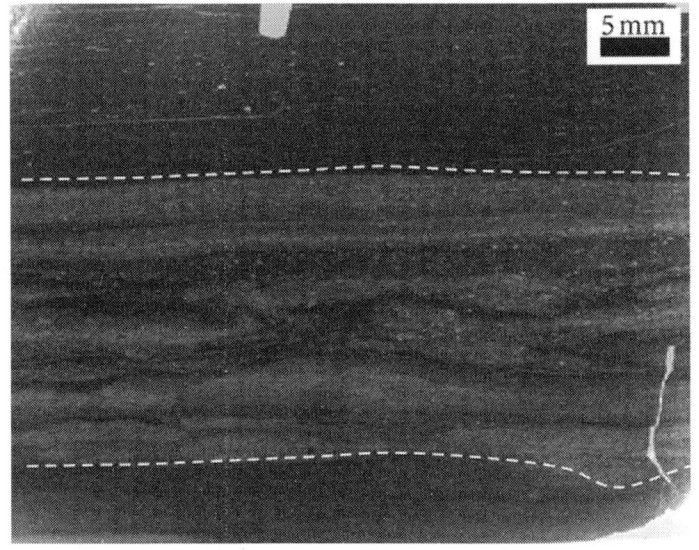

(a)

(b)

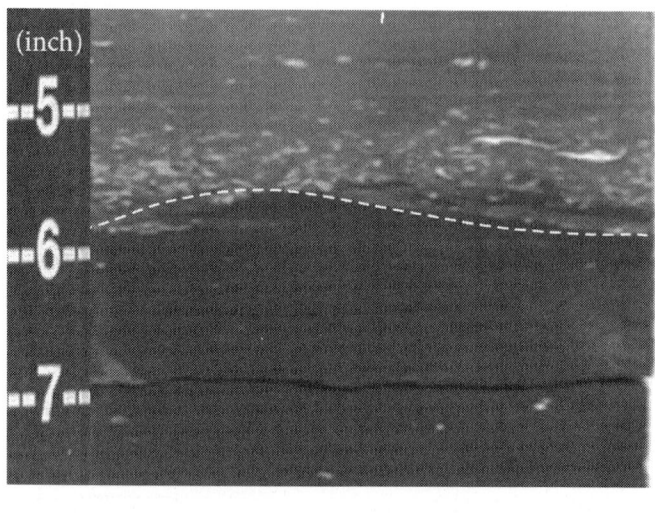

(c)

(d)

Figure 3: (a) Thin section photo shows upper sharp straight contact and lower scoured contact. (b) Thin section photo shows upper scoured contact and lower irregular contact. (c) Core photo shows lower sharp irregular contact and upper gradational contact. (d) Thin section photo shows lower contact that cut down into underlying facies.

Interpretation

The lithology changes marked by erosion surface could be associated with reworking and winnowing of the preexisting deposits as a result of prevailing tide, wave, or storm action [24]. However, these surfaces may represent a hiatus or periods of low sedimentation rate and/or sediment starved conditions. The sharp contact (Figure 3(c)) reflects an erosion event associated with a sudden sea-level fluctuation [1, 7] whereas the displacement of some surfaces into the underlying lithology is interpreted as an effect of soft sediment deformation processes (Figure 3(d)). On the contrary, the absence of soft sediment deformations below the sharp surface (Figure 3(a)) suggests the eroded shale was of firm consistency [1]. The sharp nature of the erosion surface suggests that erosion may have removed the mixed layer close to the sediment-water interface prior to deposition of the overlying unit. However, when the sharp contact marks the top of an erosion zone (Figure 3(a)), then sediment starvation was most likely and probably bottom currents that affected this location. Inversely, if the sharp nature is located at the base of the erosion zone (Figure 3(c)), then a sudden subsidence or a fall in sea level would be expected and the winnowed materials would be transported down-dip during a transgression. The organic-rich shale interval with sharp base is more erosional resistant due to the higher organic content [7]. The absence of burrowing along most of lower sharp surfaces suggests that these surfaces are not associated with a prolonged depositional hiatus and that erosion of the underlying sediments was immediately followed by, if not synchronous with, deposition of the overlying lithofacies blanket.

Shelly Laminae

Description

Shelly laminae refer to dense concentrations of thin-walled, coalesced fragments, grain to grain contacts, and broken and disarticulated shells which are common in both upper and lower Barnett Shale [9, 21]. The laminae range to a maximum thickness of 15 mm (Table 1), with most between 1 and 3 mm. Surfaces are gently curved; bedding-parallel mantled the underlying sharp irregular to straight erosion surface and blanketed gradationally by siliceous noncalcareous to calcareous mudstone facies (Figures 4(a) and 4(b)). Shelly laminae of the Barnett Shale are composed mainly of bivalves, bryozoans, brachiopods, filling branch mollusks, and echinoderms with mainly calcite filling (Figure 4(c)).

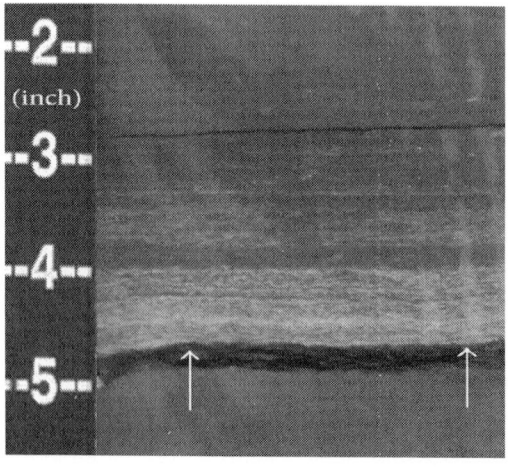

(a)

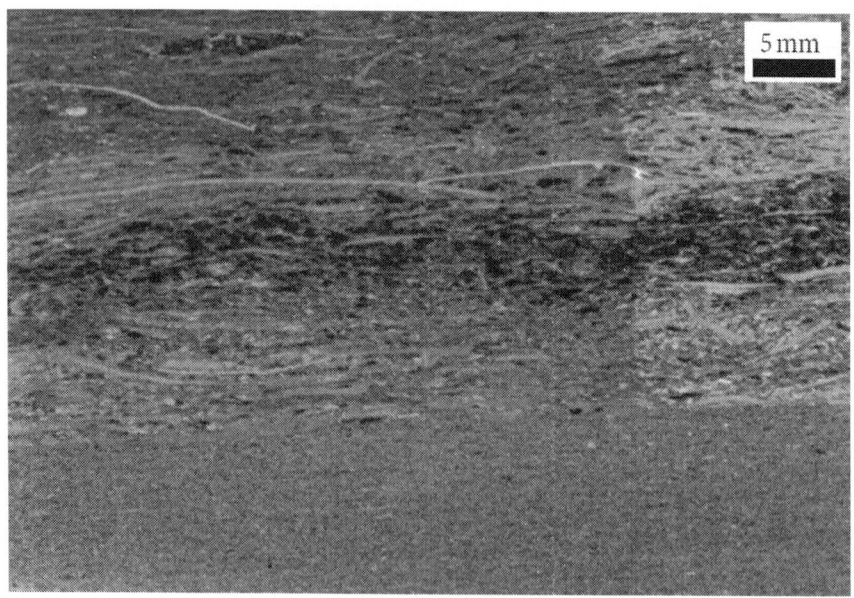

(b)

Multiscale Erosion Surfaces of the Organic-Rich Barnett Shale....   93

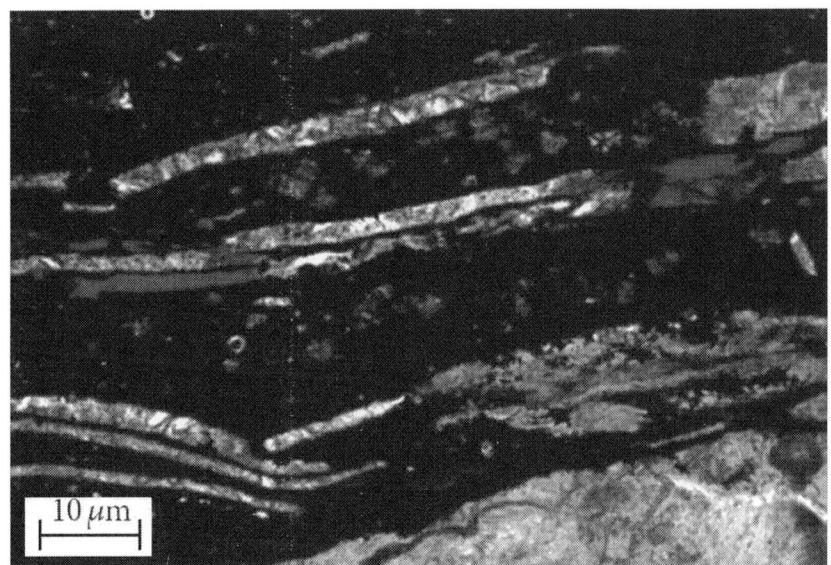

(c)

(d)

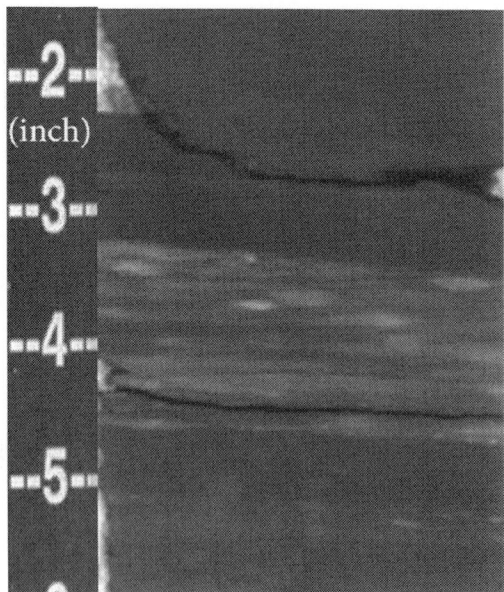

(e)

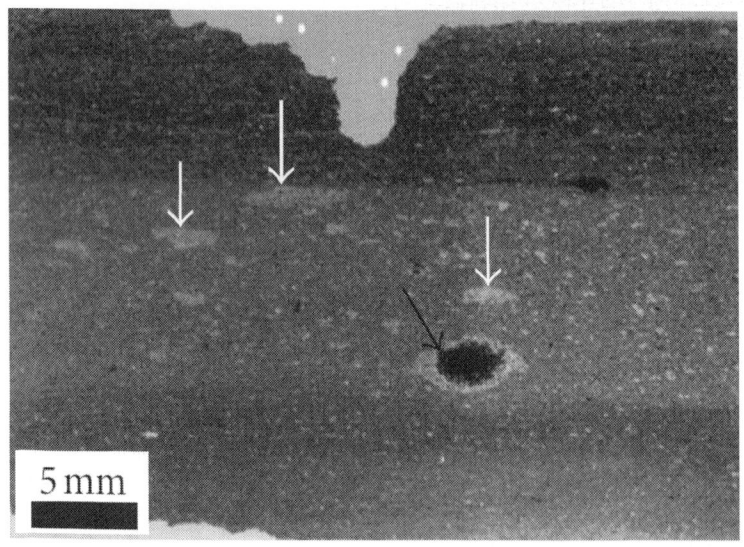

(f)

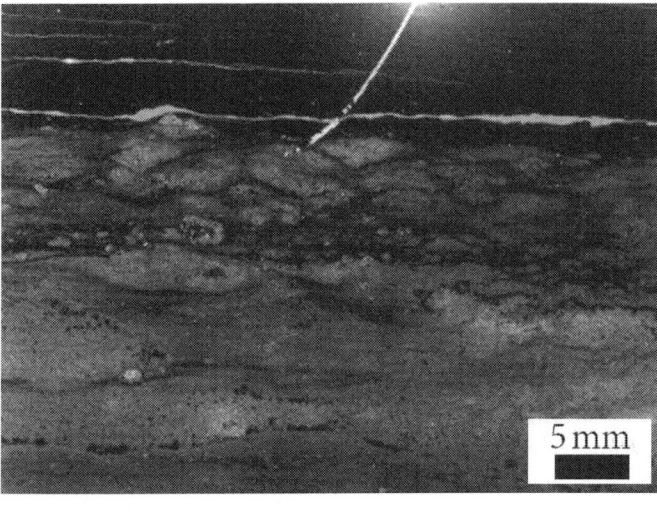

(g)

Figure 4: (a) Core photo shows highly compacted shelly laminae (arrows) with lower sharp contact and upper gradational contact. (b) Thin section photo shows lower sharp contact with overlying shelly laminae. (c) Thin section photo shows the shell fragments with calcite filling. (d) Thin section photo shows multiple shelly laminae. (e) Core photo shows shale rip-up clasts with lower and upper sharp contacts. (f) Thin section photo shows scattered shale rip-up clasts (white arrows), note the pyrite filling of some shale rip-up caslt (black arrow). (g) Thin section photo shows rounded to subrounded shale rip-up clasts with lower gradational contact and upper sharp contact.

Interpretation

Shelly laminae represents a condition of low sediment supply [25] due to the dense compacted shell fragments (Figure 4(a)); however, the rapid accumulation at the scale of such laminae should be considered [26] especially with the abundance of fossil material. The frequent occurrence of shelly laminae (Figure 4(d)) (regardless of their thickness) is interpreted as resulting either from thin compacted hiatal concentrations or multiple, varying duration episodes of erosion for the underlying deposits, that is, associated with a residual concentration of preexisting and broken fossils [26].

Shale Rip-Up Clasts

Description

Rip-up clasts are muddy chunks that swept up from the preexisting sediments and are also termed mudstone clasts, intrabasinal clasts, or clay chips. The shale rip-up clasts appear mostly without internal structures and occur as materials eroded from the underlying lithofacies (Figures 4(e) and 4(f)). These clasts exhibit different shapes and sizes: rounded to subrounded (Figure 4(g)) but flat to sharp edges; flaky-like (Figure 4(e)) is the most characteristic feature for the shale rip-up clasts in the examined samples.

Interpretation

The shale rip-up clasts occur due to the movement of coarser materials down-dip and erosion of the softer seafloor sediments. These clasts are transported and probably then suspended in basal currents and deposited when the current velocity reduced. However, shale rip-up clasts may be transported over considerable distances as evidenced by the rounded to subrounded and/or irregular surfaces (Figure 4(e)). Additionally, the flat and flaky-like shapes of shale rip-up clasts reflect compaction.

Reworked Concretions

Description

Reworked concretions are common features that are associated with the studied erosion surfaces. Typically these concretions are elongate, rounded, flat and/or irregular in shape (Figures 5(a), 5(b), and 5(c)), with a different internal grain size from the matrix ranging from a pebbly, sandy, and silty to a mud. (Figure 5(d)). Commonly concretions are light gray; however, black and dark brown concretions also occur. Interestingly, the concretions show more than one pyritized rim, indicating a multiphase origin (Figure 5(e)). The groundmass of the concretions varies between clay-rich mudstone (Figure 5(b)) and calcareous-rich mudstone (Figure 5(c)). In most cases, concretions that occur are associated with thin intervals of densely packed shells and/ or shell debris (Figure 5(c)).

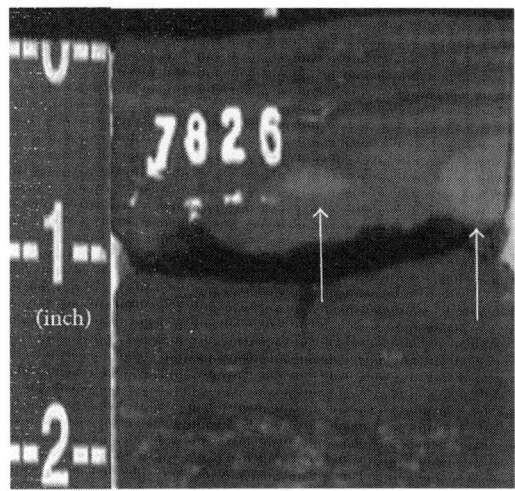

(a)

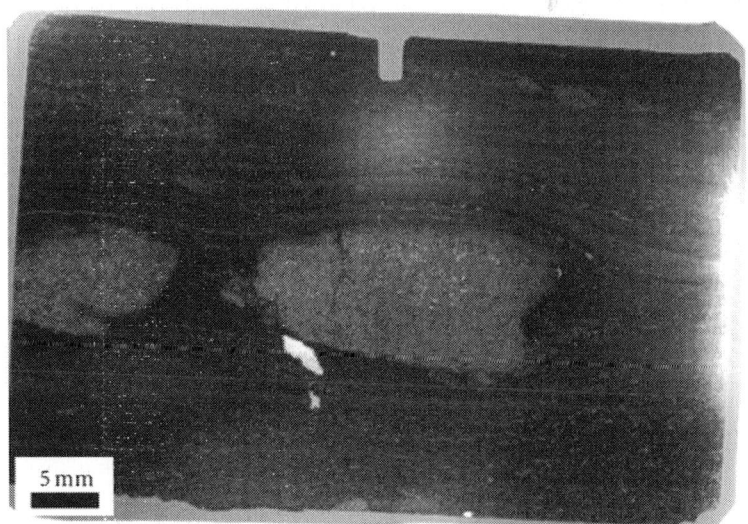

(b)

98 Deepwater Petroleum Exploration & Production

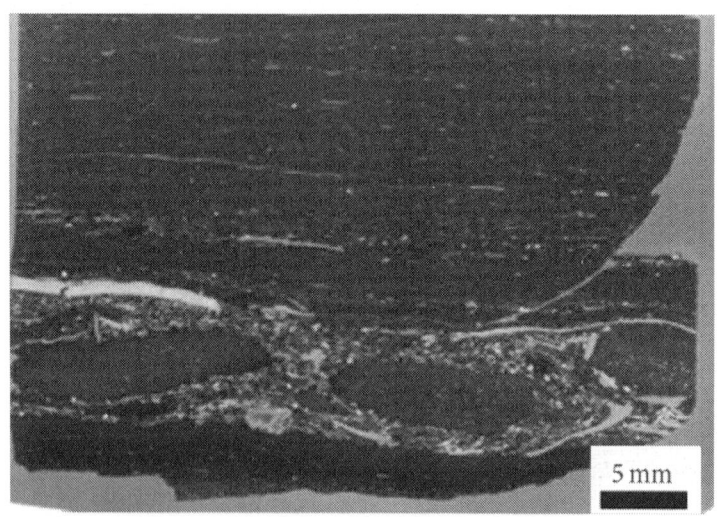

(c)

(d)

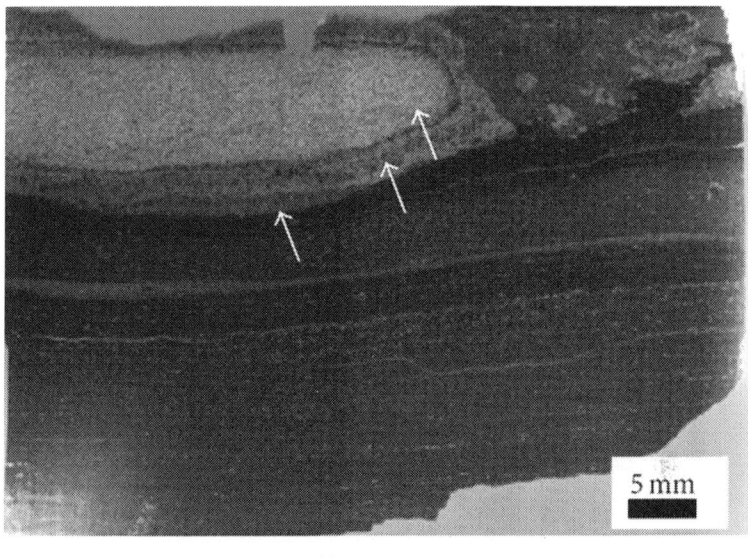

(e)

Figure 5: (a) Core photo shows reworked concretion (white arrows) associated with erosion surface with lower irregular contact and upper gradational contact. (b) Thin section photo shows large, transported subrounded concretions. (c) Thin section photo shows elongate, reworked concretions associated with irregular sharp erosional surface. (d) Thin section photo shows sandy-size reworked concretions. (e) Thin section photo shows reworked concretion with several core (white arrows) that rimmed with pyrite.

Interpretation

The roundness of concretions, different lithology from the host sediment, and the soft deformation structures in the underlying sediments suggest the reworking origin of these concretions. The occurrence of reworked concretions along erosional surfaces indicates formation at depths sufficiently shallow to be reworked by storms [23]. The dark brown color could be due to either phosphate and/or iron oxide impregnation; the latter might prevail due to the oxidation of pyrite in an oxic zone as a result of shallowing conditions. A storm event is one process that can rework concretions and is associated with a fining upward sequence afterstorm [27].

Description

The phosphatic pellets that are associated with erosion surfaces within the Barnett Shale are mostly brownish to black and range in size from few millimeters to 1 centimeter (Figure 6(a)). The pellets occur as rounded (Figure 6(b)) to subrounded, broken (Figure 6(c)), irregular, and rarely elongated with long axis parallel to bedding planes. The nuclei of these phosphatic grains may include shell fragments, detrital quartz, and glauconite grains.

(a)

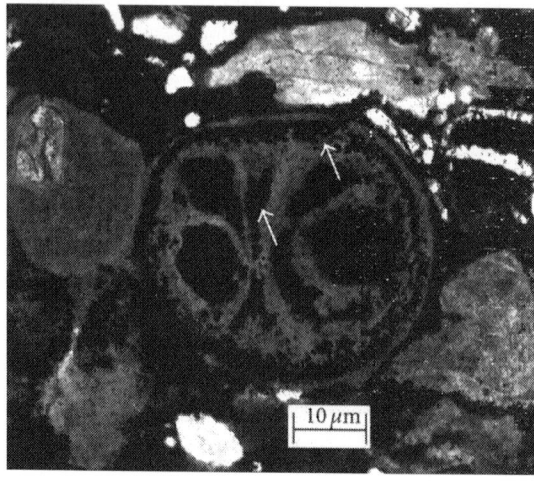

(b)

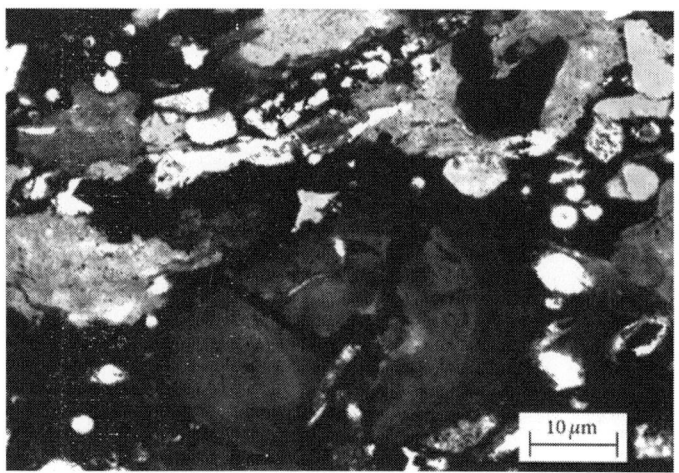

(c)

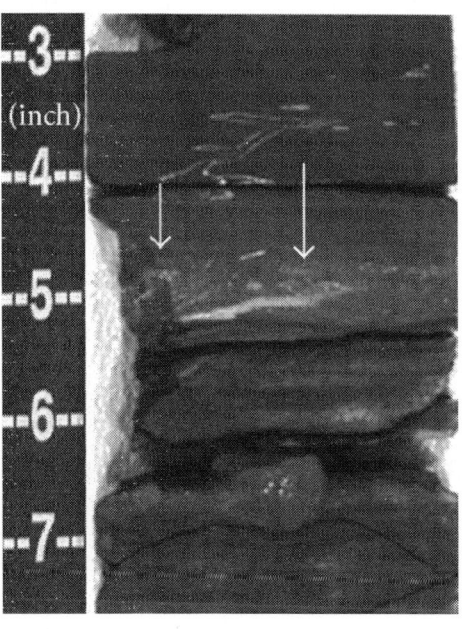

(d)

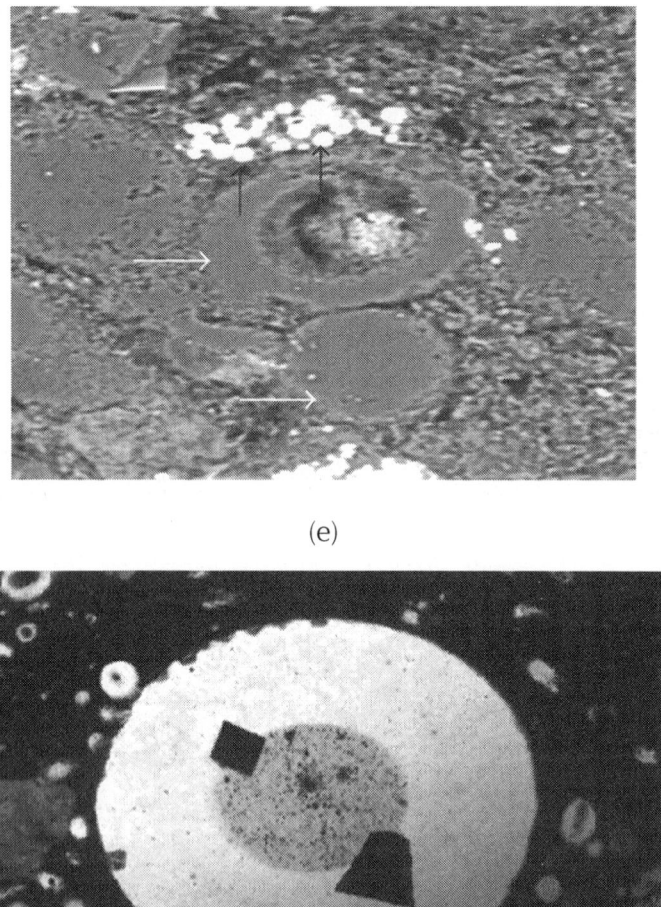

(e)

(f)

Figure 6: (a) Thin section photo shows concentration of phosphatic pellets associated with erosion surface. (b) Thin section photo shows rounded and subrounded phosphatic pellets; note the pyritic rim inside the phosphatic pellet (white arrows). (c) Thin section photo shows reworked broken phosphatic pellet associated with erosion surface. (d) Core photo shows the extensive pyrite cement of the erosion surface (white arrows). (e) Electron microprobe photo shows framboidal pyrite (black arrows) developed around sponge specula (white arrow) (photo dimensions is 527 × 527 mm). (f) Thin section photo shows fine euhedral pyrite crystal through the core of large sponge spicule associated with erosion surface.

Interpretation

The occurrence of phosphatic pellets is widely documented within the Barnett Shale [9, 28, 29], so it is one of the most important constituents associated with the erosion surfaces. Phosphatic pellets are important indicators of extensive reworking of the underlying substrate. The nucleus of these pellets shows multiple coatings through extensive reworking and transporting processes. The size, roundness, and broken edges of the phosphatic pellets, which are associated with erosion surfaces, imply that the energy conditions were sufficiently high to erode materials from the underlying sediments and agitate individual particles forming pellets.

Pyrite

Description

Pyrite occurs in erosion surfaces dominantly (Figure 6(d)) as a conspicuous accessory component disseminated in most of the phosphatic pellets (Figure 6(b)), shale rip-up clasts (Figure 4(f)), and reworked concretions (Figure 5(e)). Pyrite is present mainly in the form of framboidal pyritic pellets that may aggregate to larger pellets (Figure 6(e)) and disseminated fine euhedral crystals (Figure 6(f)).

Interpretation

The presence of pyrite is an indication of geochemical conditions during and/or after development of erosion surface [18, 30–32]. Furthermore, pyrite indicates that the erosion surfaces resulting from major sea-level drop [7]. The abundance of fine crystalline pyrite in phosphatic pellets and shale intraclasts indicate deposition from an euxinic (anoxic and sulfidic) water column [30] and hence reducing depositional environment. Pyritization of a thick zone that has taken place beneath an erosion surface may represent a significant hiatus associated with the erosion event [33]. Pyrite framboids are characteristic of early diagenesis and crystallize in the water column, settle to the bottom, and cease to grow [18, 34].

ORIGIN OF EROSION SURFACES

Origin of erosion surfaces was the focus of several works; for example, Baird et al. [35] suggested internal waves during transgression. Allersma [36] and Rine and Ginsburg [37] proposed the migration of mud waves (mud banks) as an origin for erosion surfaces whereas wave scouring in front of an advancing shoreline is also an accepted origin for the regressive surface of marine erosion and ravinement surface [38, 39]. However, Schieber [7] refers to the origin of erosion surfaces in the Chattanooga Shale, storm-induced currents. However, strong bottom-current activity could be the reason for gravel lag formation as stated by Howe et al. [40]. Additionally, storm wave action can cause erosion of the sea floor as proposed by Caron et al. [24].

The erosion events in the Barnett Shale exhibit variable relief (5–60 mm), different geometry for upper and lower bounding contacts, and a wide range of erosional features with variable grain size above and below the surface (Table 1). Clearly, several factors control such variations, including the energy conditions, rate of relative sea-level fluctuation, rate of sedimentation, sediment influx, and the lithofacies type of the underlying bed [41–45].

Although, in most cases, sediment starvation condition is the favored interpretation for development of erosion surfaces in organic-rich shale [46–49], the detrital components associated with these surfaces were likely input to distal basin from the proximal shelf as a result of wave processes eroding and reworking the underlying deposits (Figure 7(a)) [50–52]. Furthermore, the lower sharp to irregular contacts associated with rounded to sub rounded intraclasts indicate that the sea bed was being reworked (Figure 7(b)), and consequently higher-energy conditions are suggested for the development of these surfaces than previously thought [22].

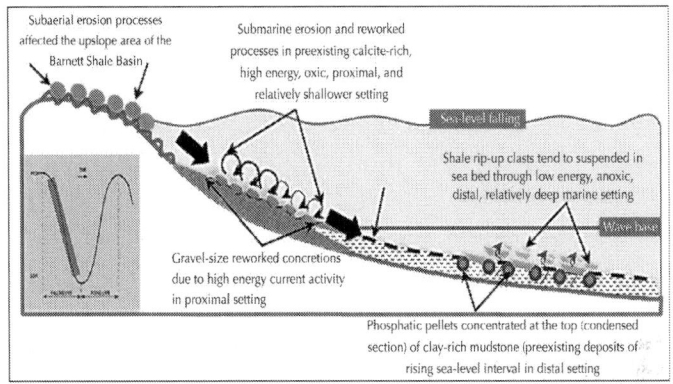

(a)

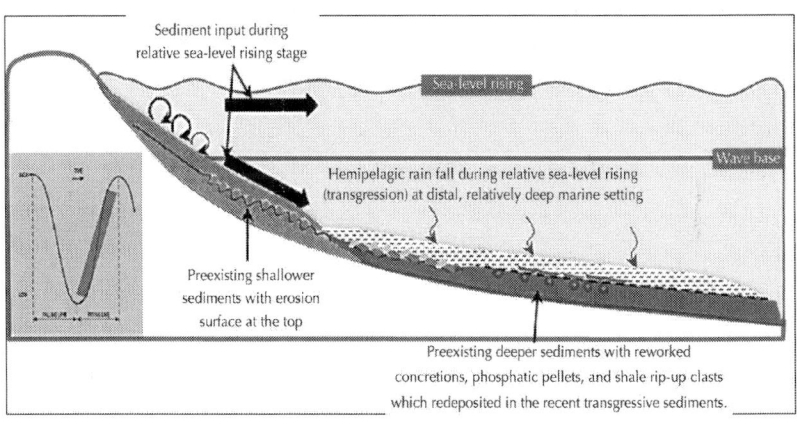

(b)

Figure 7: (a) Schematic diagram illustrates the development of erosion surface and their related features during relative sea-level fall in both proximal and distal muddy depositional settings. (b) Schematic diagram shows the transportation of the eroded and reworked materials down-dip during the relative sea-level rise to form erosion surface under the transgressive sediments.

During periods of relative sea-level fall, the action of storm currents and waves possibly caused intensive winnowing and reworking of the underlying deep marine mudstones (Figure 7(a)) [8]. The resulting erosion surfaces which may remove the underlying regressive surface of erosion have commonly been interpreted as transgressive surface of erosion [53, 54].

In some cases, the absence of erosion materials around erosion surfaces suggests either that these materials were transported to somewhere else in the basin, possibly by an unidirectional current strong enough to erode firm mud [1], or that the Barnett Shale was below the fairweather base level during a period of rapid rise of sea level which is insufficient time for sediment-water interaction [45]. However, the sharp to irregular erosion surface at the top of clay-rich shale facies (condensed section) (Figure 8(a)) may indicate that the erosion processes may have removed the expected, underlying, highstand lithofacies at this particular locality.

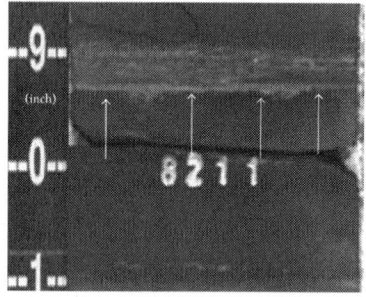

(a)

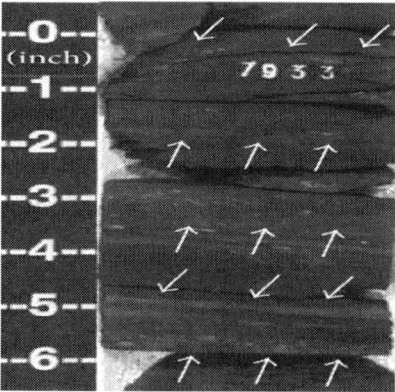

(b)

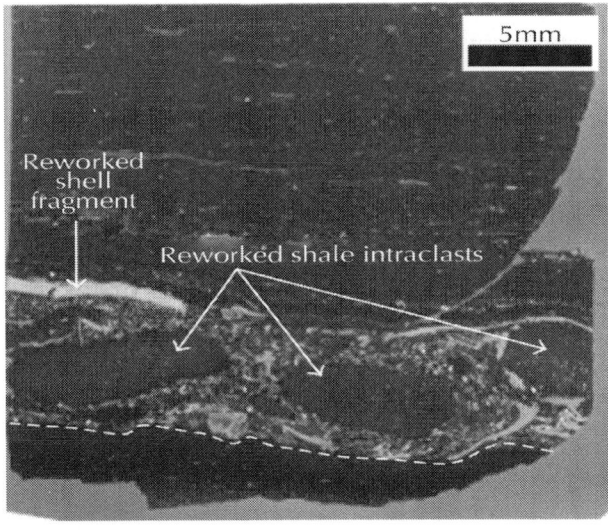

(c)

(d)

Figure 8: (a) Core photo shows sharp erosional contact at the topmost part of clay-rich mudstone facies. (b) Core photo shows multiple occurrence of erosional surfaces within less than one foot. (c) Thin section photo shows erosional surface and its related features indicated relatively proximal, shallow, and high energy muddy depositional setting (note the lower irregular contact (white dashed line), the pebbly-size reworked shale intraclasts, and reworked shell fragment). (d) Thin section photo shows erosional surface and its related features indicated relatively distal, deep, and low energy muddy depositional settings (note the lower sharp contact (white dashed line) and shale rip-up clasts).

Because the Barnett Shale is mainly a clay-rich mudstone, thus the grain size variation of the substrate has little influence on the properties of the shale as in more proximal settings. However, the high energy conditions suggested for deposition of the Barnett Shale [2, 9, 18, 21] may generate erosion surfaces with high relief [45]. Noteworthy, the frequent occurrence of erosion surfaces throughout the Barnett Shale (Figure 8(b)) suggests that the episodic storm currents and/or hyperpycnal flows are the major carriers for eroded and transported the erosional products from proximal to distal areas [2].

The development of erosion surfaces might have been influenced by water depth, sediment consistency, and slope where these conditions are expressed geologically as lateral facies changes associated with sedimentation breaks. The lateral variation from proximal (high energy preferred) to distal (low energy preferred) marine depositional settings can be indicated from the nature of erosion surfaces (Table 2). For example, the irregular nature, pebbly to sand size reworked intraclasts, reworked shelly laminae (Figure 8(c)), and the abundance of bioturbation indicate relatively shallower settings (Figures 7(a) and 7(b)), whereas sharp nature, shale rip-up clasts (Figure 8(d)), highly compacted shelly laminae, phosphatic-rich laminae, dolomitic and pyritic diagenetic cement, and the absence of bioturbation represent the relatively distal and deeper marine settings (Figures 7(a) and 7(b)).

Table 2: Characteristic features of erosion surfaces in both the proximal and the distal muddy settings

Characters	Proximal muddy settings	Distal muddy settings
Boundary nature	Irregular, scoured, cutting down, gradational to irregular	Sharp to irregular

Lithofacies	Calcareous-rich mudstone facies with dolomitic cementation	Clay-rich mudstone facies with pyritic cementation
Reworked concretions	Rounded to subrounded calcareous shale intraclasts	Rounded to subrounded phosphatic pellets
Shelly laminae	Reworked shell fragments	Compacted shelly laminae
Shale ripup clasts	Rarely occurred	Elongated, flaky with sharp edges
Energy levels	High energy enough to include reworked shale intraclasts and other components	Low energy where the shale rip-up clasts are suspended in the sea bed layer

The significance of erosion surfaces within the organic-rich Barnett Shale is not only that it aids regional sequence stratigraphic analysis [9, 29] but it also represents erosion of organic-rich muddy sediments so prediction of where these sediments have been re deposited should be considered as a future development target.

EROSION SURFACES AS MULTI-SCALE STRATIGRAPHIC SURFACES

Since the accommodation space is likely available in the distal marine environments during the time intervals of relative sea-level fall and rise, then the identification of sequential events of deposition, nondepositions and/or erosion especially within fine-grained, organic-rich shale poses a unique challenge. However, the relative sea-level fluctuations are often expressed as subtle changes in bottom water oxygen levels, influx of terrigenous materials, biogenic productivity, and deposition of highly organic-looserich condensed sections [55]. The above discussed erosional features and their associated surfaces are candidates at least in part as boundaries between different genetic types of deposits [56] but on different scales depending on changes in base-level and/or sediment supply. Subsequently, the studied erosion surfaces of the Barnett Shale reveal different strengths and/or duration which can be grouped into three different scales of stratigraphic surfaces including sequence-scale surfaces, parasequence-scale surfaces, and event-scale surfaces.

Sequence-Scale Erosion Surfaces

The sequence-scale erosional surfaces that result from a major erosional event [7] are the largest scale of the identified unconformable surfaces. This type of erosional surface is used to establish sequence stratigraphic framework for the Barnett Shale [9, 29]. Such surfaces have relatively high relief (commonly >25 mm and up to 60 mm) and have an irregular sharp boundary associated with a sandy to silty pyritic lag (Figure 9(a)). The most distinctive features of the sequence-scale erosional surfaces are the rounded to subrounded, reworked, intraclasts and highly compacted, horizontally oriented, and shelly laminae (Figure 9(a)) and (Table 3).

Table 3: Characteristic features of multi-scale erosional surfaces

Characters	Sequence-scale erosion surface	Parasequence-scale erosion surface	Event-scale erosion surface
Relief	More than 25 mm	Range between 8 and 20 mm	<5 mm
Surface nature	Irregular, scoured, cutting down	Sharp, irregular to gradational	Sharp
Lithofacies	siliceous non calcareous mudstone and Siliceous calcareous mudstone	Siliceous non calcareous mudstone and Siliceous calcareous mudstone	Clay-rich mudstone facies
Reworked concretions	Rounded to subrounded shale intraclasts	Subrounded phosphatic pellets	Calcareous concretions
Shelly laminae	Highly compacted and horizontally oriented shell fragments	Traces of thin disconnected shelly laminae	Rarely occurred
Shale rip-up clasts	Rarely occurred	Common	Rarely occurred
Spectral gamma ray pattern	Abrupt change (decreasing and increasing)	Marking the end of upward increasing or decreasing spectral Gamma Ray pattern	Does not Apply

Energy levels	High energy enough to reworked shale intraclasts and other components	Low energy where the shale rip-up clasts suspended in the sea bed layer	Low bottom energy levels

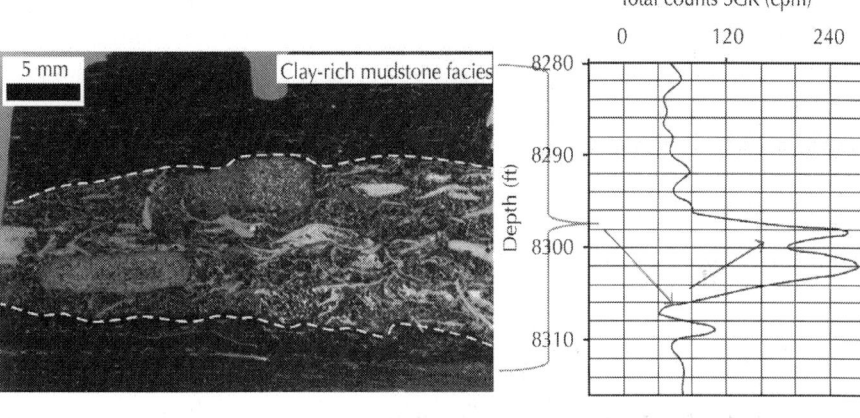

(a)

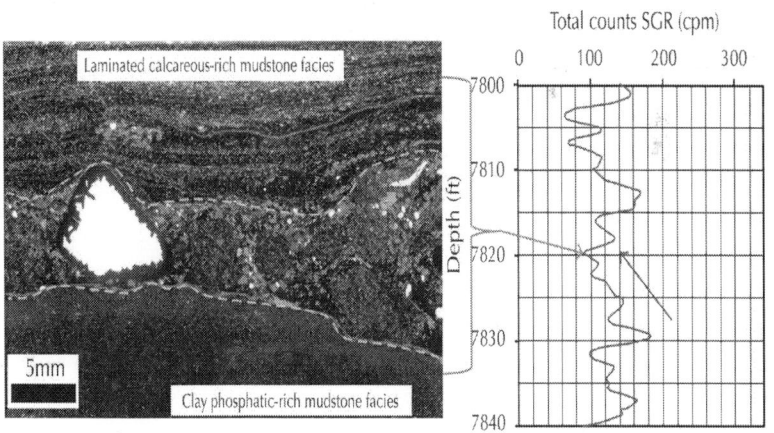

(b)

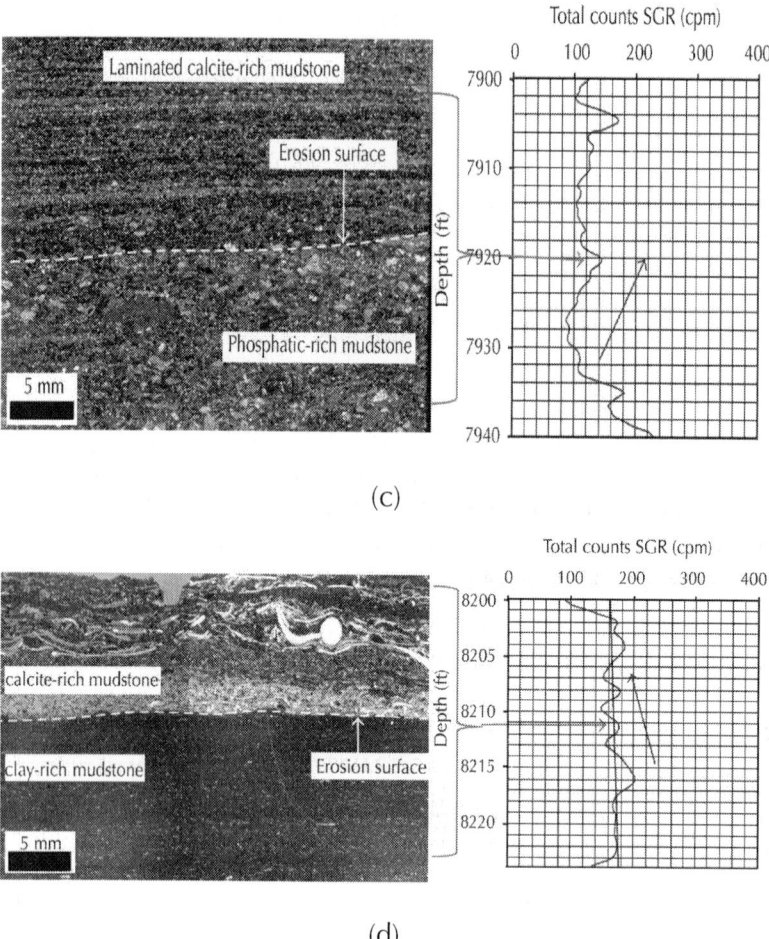

Figure 9: (a) Thin section photo shows sequence-scale erosion surface with irregular nature, pyritic reworked intraclasts, and reworked shell fragments with the opposite spectral gamma ray response that shows the abrupt upward increasing above the erosion surface. (b) Thin section photo shows irregular sequence-scale erosion surfaces; note the underlying clay phosphatic-rich facies and the overlying laminated calcareous facies. The opposite log is the spectral gamma ray value of this interval showing the coincide upward-decreasing pattern. (c) Thin section photo shows parasequence-scale erosion surface with underlying phosphatic-rich mudstone facies with the opposite Spectral Gamma Ray response that shows the upward decreasing above the erosion surface. (d) Thin section photo shows irregular parasequence-scale erosion surfaces; note the underlying clay phosphatic-rich facies and the

overlying laminated calcareous facies. The opposite log is the spectral gamma ray value of this interval showing coinciding upward-decreasing pattern.

The lithology underlying these surfaces are commonly siliceous noncalcareous mudstone (Figure 9(b)) or siliceous calcareous mudstone, whereas the overlying lithofacies show an increase in calcite content and an increasing grain size of the detrital components (Figure 9(b)). This lithological variation bounding the sequence-scale erosional surfaces reflect a lateral facies changes from clay rich to calcite rich and the vertical change from relatively deeper to relatively shallower facies which is interpreted as a major change in depositional conditions.

The sequence-scale erosional surfaces are characterized by an abrupt change on gamma ray log (Figures 9(a)–9(d)). In such a case, when the erosion surface coms at the base of upward increasing spectral gamma ray (SGR) interval (Figure 9(a)), then it is interpreted as transgressive surface of erosion (TSE) [9]. Alternatively, the erosion surface associated with abrupt decreasing in SGR values (Figure 9(b)) is a sequence boundary [21].

Parasequence-Scale Erosion Surfaces

The erosion surfaces at this scale represent a major shift in depositional settings accompanied by corresponding changes in environmental energy and sediment supply during either sea-level rise or fall [57]. Examples of changes in depositional trends include the change from sedimentation to erosion and/or starvation and vice versa and also the change from a regression to a transgression and vice versa [58].

The parasequences-scale erosion surfaces are characterized by medium relief (ranging between 8 and 20 mm) with sharp to irregular contacts underlying with phosphatic-rich to clay-rich non calcareous mudstone facies (Figure 9(c)) of relatively deeper depositional environment and overlain with laminated calcareous-rich mudstone facies of relatively shallower depositional settings [9, 18, 21]. Additionally, this type of erosion surfaces represents the turnaround point on SGR curve as a result of the lithofacies change from clay, phosphatic-rich mudstone below to siliceous calcareous mudstone above these surfaces (Figure 9(c)) (Table 3). The parasequence-scale erosion surface may consistent with a marine flooding surface and/

or maximum flooding surface which results from an abrupt increase in water depth [59, 60] when it is associated with shale rip-up clasts, traces of compacted and horizontally oriented shell fragments (Figure 9(d)) as well as diagenetic dolomite and pyrite.

Event-Scale Erosion Surfaces

These are the smallest scale types of erosional surfaces that mark changes in sedimentation regimes and may reflect minor or short-duration erosional events [7]. Such events are commonly distinguished by flat, low relief (<5 mm), sharp contacts, and shale on shale boundary with minimal evidence of erosion (Figure 10(a)) (Table3). The sharp nature of these erosional surfaces is attributed to abrupt lithological changes and the presence of calcareous concretions at surfaces as a result of sediment starvation or a low sedimentation rate reflecting low depositional energy [61]. The lithofacies change is markedly around this type of erosional surface with subsequent deposition of siliceous non calcareous mudstone (Figure 10(b)) to a siliceous calcareous mudstone (Figure 10(c)). Generally, the high frequency and diversity of erosion surfaces through the Barnett Shale give a unique view into the short-duration stratigraphic intervals that were previously difficult to detect in fine grained rock.

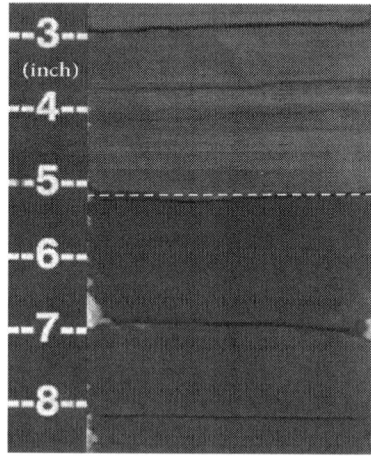

(a)

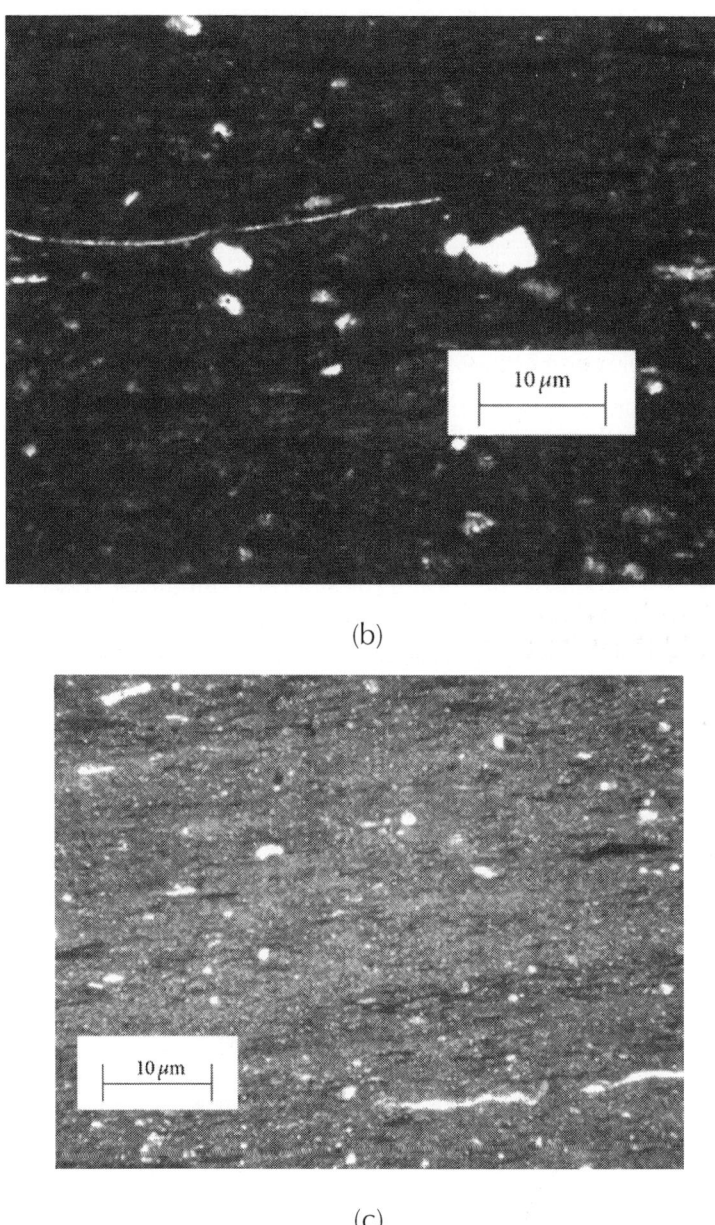

(b)

(c)

Figure 10: (a) Core photo shows an event-scale erosion surface with sharp nature (dashed line) between laminated black shale and an overlying dark laminated gray shale bed. (b) Thin section photo shows siliceous noncalcareous mudstone facies below the event-scale erosion surface in (a). (c) Thin section photo shows siliceous calcareous mudstone facies above the event-scale erosion surface in (a).

Summary and Conclusion

The erosion surfaces are direct indications of major environmental changes that may include nondeposition and/or erosional events. The erosion surfaces are relatively abundant throughout the Barnett Shale which punctuate the stacking pattern and erode previously deposited strata to an unknown extent.

The erosion surfaces in the organic-rich Barnett Shale exhibit variable relief (5–60 mm) with different geometry of upper and lower bounding contacts and a wide range of variable features including shelly laminae, shale rip-up clasts, reworked intraclasts, phosphatic pellets, and diagenetic minerals (dolomite and pyrite). Several factors control such variations, including the energy conditions at deposition, rate of relative sea-level fluctuation, rate of sedimentation, sediment influx, and the lithofacies type of the underlying and the overlying beds. The alternative occurrence of erosion surfaces throughout the Barnett Shale suggests episodic storm currents which transport eroded particles and intraclasts from proximal to distal areas.

The erosional features and associated surfaces are in part boundaries between different genetic types of deposits but on different scales reflecting the dependence on either base-level and/or sediment supply. Accordingly, the studied erosional surfaces of the Barnett Shale can be grouped into three different scales of sequence stratigraphic surfaces: sequence-scale surfaces, parasequence-scale surfaces, and event-scale surfaces.

ACKNOWLEDGMENTS

The author is thankful to the ConocoPhillips School of Geology and Geophysics, University of Oklahoma, where the labwork of this research has been conducted. The author is also thankful to the Faculty of Petroleum and Mining Engineering, Suez University, for providing the sabbatical leave. The constructive reviews and editorial comments of Roger M. Slatt, Gungoll Family Chair Professor of Petroleum Geology and Geophysics and Director of Reservoir Characterization Institute, University of Oklahoma, are gratefully acknowledged.

REFERENCES

1. J. Schieber, "Evidence for high-energy events and shallow-water deposition in the Chattanooga Shale, Devonian, central Tennessee, USA," Sedimentary Geology, vol. 93, no. 3-4, pp. 193–208, 1994.
2. M. O. Abouelresh and R. M. Slatt, "Shale depositional processes: example from the paleozoic Barnett Shale, Fort Worth Basin, Texas, USA," Central European Journal of Geoscience, vol. 3, no. 4, pp. 398–409, 2011.
3. W. J. Kennedy and H. C. Klinger, "Hiatus-concretions and hardground horizons in the Cretaceous of Zululand," Paleontology, pp. 539–549, 1972.
4. W. J. Kennedy and R. E. Garrison, "Morphology and genesis of nodular chalks and hardgrounds in the Upper Cretaceous of Southern England," Sedimentology, vol. 22, pp. 311–386, 1975.
5. W. J. Kennedy and R. E. Garrison, "Morphology and genesis of nodular phosphates in the Cenomanian Glauconitic Marl of southeast England," Lethaia, vol. 8, pp. 339–360, 1975.
6. G. C. Baird, "Pebbly phosphorites in shale, a key to recognition of a widespread submarine discontinuity in the Middle Devonian of New York," Journal of Sedimentary Petrology, vol. 2, pp. 545–555, 1978.
7. J. Schieber, "Sedimentary features indicating erosion, condensation, and hiatuses in the Chattanooga Shale of Central Tennessee: relevance for sedimentary and stratigraphic evolution," in Shales and Mudstones, J. Schieber, W. Zimmerle, and P. Sethi, Eds., vol. 1, pp. 187–215, 1998.
8. D. Nummedal and D. J. P. Swift, "Transgressive stratigraphy at sequence-bounding unconformities: some principles derived from Holocene and Cretaceous examples," in Sea-Level Fluctuation and Coastal Evolution, D. Nummedal, O. H. Pilkey, and J. D. Howard, Eds., vol. 42, pp. 358–370, SEPM Special Publication, 1987.

9. M. O. Abouelresh and R. M. Slatt, "Lithofacies and sequence stratigraphy of the Barnett Shale in east-central Fort Worth Basin, Texas," AAPG Bulletin, vol. 96, no. 1, pp. 1–22, 2012.
10. S. L. Montgomery, D. M. Jarvie, K. A. Bowker, and R. M. Pollastro, "Mississippian Barnett Shale, Fort Worth basin, north-central Texas: Gas-shale play with multi-trillion cubic foot potential," AAPG Bulletin, vol. 89, no. 2, pp. 155–175, 2005.
11. P. T. Flawn, A. Goldstein Jr., P. B. King, and C. E. Weaver, "The Ouachita system: University of Texas, Bureau of Economic Geology," Report 6120, 1961.
12. J. L. Walper, "Plate tectonic evolution of the Fort Worth Basin," in Petroleum Geology of the Fort Worth Basin and Bend Arch Area, C. A. Martin, Ed., pp. 237–251, Dallas Geological Society, Dallas, Tex, USA, 1982.
13. J. K. Arbenz, "Structure framework of the ouachita mountains," in Stratigraphic and Structural Evolution of the Ouachita Mountains and Arkoma Basin, Southeastern Oklahoma and West-Central Arkansas: Application To Petroleum Exploration, N. H. Suneson, Ed., pp. 4–40, Oklahoma Geological Survey, Circular, 2009.
14. J. D. Henry, "Stratigraphy of the Barnett Shale (Mississippian) and associated reefs in the northern Fort Worth Basin," in Petroleum Geology of the Fort Worth Basin and Bend Arch Area, C. A. Martin, Ed., pp. 157–178, Dallas Geological Society, Dallas, Tex, USA, 1982.
15. J. W. Flippin, "The stratigraphy, structure, and economic aspects of the Paleozoic strata in Erath County, north central Texas," in Petroleum Geology of the Fort Worth Basin and Bend Arch Area, C. A. Martin, Ed., pp. 129–155, Dallas Geological Society, Dallas, Tex, USA, 1982.
16. R. S. Kier, L. F. Brown, and E. F. McBride, The Mississippian and Pennsylvanian (Carboniferous) Systems in the United States: Texas, vol. 14, Bureau of Economic Geology, University of Texas at Austin, Geological Circular, Austin, Texas, 1980.
17. W. J. Mapel, R. B. Johnson, G. O. Bachman, and K. L. Varnes, "Southern midcontinent and southern Rocky Mountains region," in Paleotectonic Investigations of the Mississippian System in the United States, L. C. Craig and C. W. Connor, Eds., vol. 1010, pp. 161–187, Geological Survey Professional Paper, 1979.

18. R. G. Loucks and S. C. Ruppel, "Mississippian Barnett Shale: Lithofacies and depositional setting of a deep-water shale-gas succession in the Fort Worth Basin, Texas," AAPG Bulletin, vol. 91, no. 4, pp. 579–601, 2007.
19. H. D. Rowe, R. G. Loucks, S. C. Ruppel, and S. M. Rimmer, "Mississippian Barnett Formation, Fort Worth Basin, Texas: Bulk geochemical inferences and Mo-TOC constraints on the severity of hydrographic restriction," Chemical Geology, vol. 257, no. 1-2, pp. 16–25, 2008.
20. B. U. Haq and S. R. Schutter, "A chronology of paleozoic sea-level changes," Science, vol. 322, no. 5898, pp. 64–68, 2008.
21. P. Singh, Lithofacies and sequence-stratigraphic framework of the Barnett Shale, northeast Texas [Ph.D. thesis], University of Oklahoma, Norman, Oklahoma, 2008.
22. J. Macquaker, "Micro-Textural Analyses of Fine-Grained Sediments and the Roles that Advective Sediment Transport and Suspension Settling Processes Play in the Deposition of Fine-Grained Organic Carbon-Rich Sediments. OR Just How Shaky are the Current Depositional Models that Seek to Explain the Origin of Source Rocks / shale Gas Reservoirs?" in Proceedings of the Critical assessment of shale resource plays (ex. abs.), American Association Of Petroleum Geologists/Society of Economic Geologists/Society of Petroleum Geologists/ Society of Petrophysicists and Well-Log Analysis Hedberg Conference, Austin, Texas, USA, December 2010.
23. P. E. Potter, J. B. Maynard, and P. J. Depetris, Mud and Mudstones: Introduction and Overview, Springer, Berlin, Germany, 2005.
24. V. Caron, C. S. Nelson, and P. J. J. Kamp, "Transgressive surfaces of erosion as sequence boundary markers in cool-water shelf carbonates," Sedimentary Geology, vol. 164, no. 3-4, pp. 179–189, 2004.
25. A. P. Heward, "A review of wave-dominated clastic shoreline deposits," Earth Science Reviews, vol. 17, no. 3, pp. 223–276, 1981.
26. S. M. Kidwell, "The stratigraphy of shell concentrations," in Taphonomy, Releasing the Data Locked in the Fossil Record, P. A. Allison and D. E. G. Briggs, Eds., pp. 211–290, Plenum, New York, NY, USA, 1991.

27. F. T. Fursich, W. Oschmann, I. B. Singh, and A. K. Jaitly, "Hardgrounds, reworked concretion levels and condensed horizons in the Jurassic of western India: their significance for basin analysis," Journal of Geological Society, vol. 149, no. 3, pp. 313–331, 1992.
28. J. J. Hickey and B. Henk, "Lithofacies summary of the Mississippian Barnett Shale, Mitchell 2 T.P. Sims well, Wise County, Texas," AAPG Bulletin, vol. 91, no. 4, pp. 437–443, 2007.
29. P. Singh, R. M. Slatt, G. Borges et al., "Reservoir characterization of unconventional gas shale reservoirs: example from the Barnett Shale, Texas, U.S.A," Oklahoma City Geological Society Shale Shaker, vol. 60, no. 1, pp. 15–31, 2009.
30. R. T. Wilkin, H. L. Barnes, and S. L. Brantley, "The size distribution of framboidal pyrite in modern sediments: an indicator of redox conditions," Geochimica et Cosmochimica Acta, vol. 60, no. 20, pp. 3897–3912, 1996.
31. R. T. Wilkin, M. A. Arthur, and W. E. Dean, "History of water-column anoxia in the Black Sea indicated by pyrite framboid size distributions," Earth and Planetary Science Letters, vol. 148, no. 3-4, pp. 517–525, 1997.
32. D. Bond, P. B. Wignall, and G. Racki, "Extent and duration of marine anoxia during the Frasnian-Famennian (Late Devonian) mass extinction in Poland, Germany, Austria and France," Geological Magazine, vol. 141, no. 2, pp. 173–193, 2004.
33. K. L. Bann, C. R. Fielding, J. A. MacEachern, and S. C. Tye, "Differentiation of estuarine and offshore marine deposits using integrated ichnology and sedimentology: Permian Pebbly Beach Formation, Sydney Basin, Australia," in The Application of Ichnology To Palaeoenvironmental and Stratigraphic Analysis, D. McIlroy, Ed., vol. 228, pp. 179–211, Geological Society, London, UK, 2004.
34. J. P. Herbin, C. Muller, J. R. Geyssant, F. Melieres, I. E. Penn, and Y. Group, "Variation of the distribution of organic matter within a transgressive system tract: Kimmeridge Clay (Jurassic), England," in Source Rocks in a Sequence Stratigraphic Framework, B. J. Katz and L. M. Pratt, Eds., vol. 37, pp. 67–100, The American Association of Petroleum Geologists, 1993.
35. G. C. Baird, C. E. Brett, and W. T. Kirchgasser, "Genesis of

black shale-roofed discontinuities in the Devonian Genesee Formation, Western New York State," in Devonian of the World, N. J. McMillan, A. F. Embry, and D. J. Glass, Eds., vol. 2, pp. 357–375, Canadian Society of Petroleum Geologists, Calgary, Canada, 1988.

36. E. Allersma, "Mud on the oceanic shelf of Guiana," in Proceedings of the Symposium on Investigation and Resources of the Caribbean Sea and Adjacent Regions, pp. 193–203, UNESCO, Paris, France.

37. J. M. Rine and R. N. Ginsburg, "Depositional facies of a mud shoreface in Suriname, South America: a mud analogue to sandy, shallow-marine deposits," Journal of Sedimentary Petrology, vol. 55, no. 5, pp. 633–652, 1985.

38. A. G. Plint, "Sharp-based shoreface sequences and "offshore bars" in the Cardium Formation of Alberta, their relationship to relative changes in sea level," in Sea Level Changes: An Integrated Approach, C. K. Wilgus, B. S. Hastings, C. G. St. C. Kendall, H. W. Posamentier, C. A. Ross, and J. C. Van Wagoner, Eds., vol. 42, pp. 357–370, SEPM Special Publication, 1988.

39. D. Nummedal, G. W. Riley, and P. L. Templet, "High-resolution sequence architecture: a chronostratigraphic model based on equilibrium profile studies," in Sequence Stratigraphy and Facies Association, H. Posamentier, C. P. Summerhayes, B. U. Haq, and C. P. Allen, Eds., vol. 18, pp. 55–68, SEPM Special Publication, 1993.

40. J. A. Howe, M. S. Stoker, and K. J. Woolfe, "Deep-marine seabed erosion and gravel lags in the northwestern Rockall Trough, North Atlantic Ocean," Journal of the Geological Society, vol. 158, no. 3, pp. 427–438, 2001.

41. J. A. MacEachern, B. A. Zaitlin, and S. G. Pemberton, "High-resolution sequence stratigraphy of early transgressive deposits, Viking Formation, Joffre Field, Alberta, Canada," AAPG Bulletin, vol. 82, no. 5 A, pp. 729–756, 1998.

42. S. A. J. Pattison, "Sequence stratigraphic significance of sharp-based lowstand shoreface deposits, Kenilworth Member, Book Cliffs, Utah," AAPG Bulletin, vol. 79, no. 3, pp. 444–462, 1995.

43. S. M. Kidwell and P. J. Brenchley, "Patterns in bioclastic accumulation through the Phanerozoic: changes in input or in destruction?" Geology, vol. 22, no. 12, pp. 1139–1143, 1994.

44. R. G. Walker, "Sedimentary and tectonic origin of a transgressive surface of erosion: Viking Formation, Alberta, Canada," Journal of Sedimentary Research B, vol. 65, pp. 209–221, 1995.
45. I. G. Hwang and P. L. Heller, "Anatomy of a transgressive lag: Panther Tongue Sandstone, Star Point Formation, central Utah," Sedimentology, vol. 49, no. 5, pp. 977–999, 2002.
46. A. Hallam, "Eustatic cycles in the Jurassic," Palaeogeography, Palaeoclimatology, Palaeoecology, vol. 23, pp. 1–32, 1978.
47. M. Elrick and J. F. Read, "Cyclic ramp-to-basin carbonate deposits, Lower Mississippian, Wyoming and Montana: a combined field and computer modeling study," Journal of Sedimentary Petrology, vol. 61, no. 7, pp. 1194–1224, 1991.
48. P. B. Wignall and J. R. . Maynard, "The sequence stratigraphy of transgressive black shales," in Source Rocks in a Sequence stratigraphic framework, B. J. Katz and L. Pratt, Eds., vol. 37, pp. 35–47, The American Association of Petroleum Geologists, Studies in Geology, 1993.
49. P. B. Wignall, Black Shales, Clarendon Press Oxford, 1994.
50. D. J. P. Swift and J. A. Thorne, "Sedimentation on continental margins I: a general model for shelf sedimentation," in Shelf Sand and Sandstone Bodies, D. J. P. Swift, G. F. Oertel, R. W. Tillman, and J. A. Thorne, Eds., vol. 14, pp. 3–31, International Association of Sedimentologists, 1991.
51. D. J. P. Swift, S. Phillips, and J. A. Thorne, "Sedimentation on continental margins I: a general model for shelf sedimentation," in Shelf Sand and Sandstone Bodies, D. J. P. Swift, G. F. Oertel, R. W. Tillman, and J. A. Thorne, Eds., vol. 14, pp. 153–187, International Association of Sedimentologists, 1991.
52. R. G. Walker and A. G. Plint, "Wave- and storm-dominated shallow marine systems," in Facies Models: Response To Sea Level Change, R. G. Walker and N. P. James, Eds., pp. 219–238, Geological Association of Canada, 1992.
53. J. P. Bhattarcharya, "The expression and interpretation of marine flooding surfaces and erosional surfaces in core, examples from the upper Cretaceous Dunvegan Formation, Alberta Foreland basin, Canada," in Sequence Stratigraphy and Facies Associations, H. W. Posamentier, C. P. Summerhayes, B. U. Haq, and G. P. Allen,

Eds., vol. 18 of International Association of Sedimentologists, Special Publication, pp. 125–160, Blackwell Publishing, 1993.

54. A. F. Embry, "Sequence boundaries and sequence hierarchies: problems and proposals," in Sequence Stratigraphy: Advances and Applications for Exploration and Production in Northwest Europe. Stavanger, R. J. Steel, V. L. Felt, E. P. Johannesen, and C. Mathieu, Eds., pp. 1–11, Elsevier, Amsterdam, The Netherlands, 1995.

55. K. M. Husley, Lithofacies characterization and sequence stratigraphic framework for some gas-bearing shales within the Horn River Basin, Northeastern British Colombia [M.S. thesis], University of Oklahoma, Norman, Oklahoma, 2011.

56. O. Catuneanu, V. Abreu, J. P. Bhattacharya, et al., "Toward the standardization of the sequence stratigraphy," Earth-Science Reviews, vol. 92, no. 1-2, pp. 1–33, 2009.

57. O. Catuneanu, Principles of Sequence Stratigraphy, Elsevier, Amsterdam, The Netherlands, 2006.

58. A. Embry, E. Johannessen, D. Owen, B. Beauchamp, and P. Gianolla, "Sequence stratigraphy as a "Concrete" stratigraphic discipline," Report of the ISSC Task Group on Sequence Stratigraphy, 2007.

59. K. M. Bohacs and J. R. Schwalbach, "Sequence stratigraphy of fine-grained rocks with special reference to the Monterey Formation," in Sequence Stratigraphy in Fine-Grained Rocks: Examples From the Monterey Formation, J. R. Schwalbach and K. M. Bohacs, Eds., vol. 70, Pacific Section SEPM Guidebook, 1992.

60. M. R. Slatt, Stratigraphic Reservoir Characterization for Petroleum Geologists, Geophysicists, and Engineers, Elsevier, 2006.

61. R. Raiswell and D. E. Canfield, "Sources of iron for pyrite formation in marine sediments," The American Journal of Science, vol. 298, no. 3, pp. 219–245, 1998.

Chapter 5

Flexible Riser Monitoring using Hybrid Magnetic/Optical Strain Gage Techniques through RLS Adaptive Filtering

Daniel Pipa, Sérgio Morikawa, Gustavo Pires, Claudio Camerini, and JoãoMárcio Santos

Materials, Equipments and Corrosion Department (TMEC), Petrobras' Research and Development Center (CENPES), Av. Horácio Macedo, 950. Cidade Universitária, 21941-915 Rio de Janeiro, RJ, Brazil

ABSTRACT

Flexible riser is a class of flexible pipes which is used to connect subsea pipelines to floating offshore installations, such as FPSOs (floating production/storage/off-loading unit) and SS (semisubmersible)

platforms, in oil and gas production. Flexible risers are multilayered pipes typically comprising an inner flexible metal carcass surrounded by polymer layers and spiral wound steel ligaments, also referred to as armor wires. Since these armor wires are made of steel, their magnetic properties are sensitive to the stress they are subjected to. By measuring their magnetic properties in a nonintrusive manner, it is possible to compare the stress in the armor wires, thus allowing the identification of damaged ones. However, one encounters several sources of noise when measuring electromagnetic properties contactlessly, such as movement between specimen and probe, and magnetic noise. This paper describes the development of a new technique for automatic monitoring of armor layers of flexible risers. The proposed approach aims to minimize these current uncertainties by combining electromagnetic measurements with optical strain gage data through a recursive least squares (RLSs) adaptive filter.

INTRODUCTION

Flexible risers are an important component of offshore production systems of oil and gas. They are used to link subsea pipelines to floating installations, such as FPSOs (floating production/storage/off-loading unit). Flexible risers have been one of the preferred deepwater riser solutions in many regions of the world due to their good dynamic behavior and reliability [1].

Petrobras is a Brazilian multinational petroleum company whose businesses include oil and gas exploration, production, transportation, refining, and distribution. Since most Brazilian oil reserves are located offshore and often under deepwater, Petrobras oil production is highly dependent on platforms and offshore equipments such as flexible risers. Integrity management of flexible risers is essential to ensure the safe operation of a production unit.

The main failure mode of flexible risers, when operating in deep waters, occurs at the riser's top section close to end fitting due to fatigue in tensile armor wires. It is known, however, that riser failure only happens after the rupture of a significant number of wires. Therefore, the structural integrity of a riser-end fitting connection may be assessed through the monitoring of wire rupture. Close to the end fitting, wires are subjected to tensile stress at 30% to 50% of yield point. Considering that rupture reduces stress to zero, the structural integrity of end fitting connection may be also assessed through monitoring tensile armor wire stress [2].

By identifying an unstressed wire among stressed ones, it is possible to infer that the remnant wires are subjected to higher loads than their operational design. This implies that riser integrity is uncertain and insecure. MAPS-FR(MAPS is a registered trade mark of MAPS Technology Ltd.) is an equipment capable of magnetically comparing the stress in the wires through the polymer layers, thus indicating broken wires and assessing riser integrity.

However, one encounters several sources of noise when measuring electromagnetic properties contactlessly, such as movement between specimen and probe, and magnetic noise. This paper describes the development of a new technique for automatic monitoring of armor layers of flexible risers. The proposed approach aims to minimize these current uncertainties by combining electromagnetic measurements with optical strain gage data through a recursive least squares (RLSs) adaptive filtering technique.

This paper is organized as follows. Section 2 introduces the flexible riser and comments its main failure modes. Section 3 presents the proposed method as well as each of its components. Finally, Section 4 presents the results obtained in a laboratory trial, which attests the potential of the method. A conclusion is drawn in Section 5. "Flexible pipe" is a general term and denotes a type of pipe, whereas "flexible riser" designates the vertical segment of a pipe which is usually connected to an offshore production unit.

This article deals with signal processing algorithms, rather than physical phenomena underlying the correspondence between mechanical load in ferromagnetic materials and their electromagnetic properties. The idea is to show that this relation does not need be fully determined and understood if one uses a global load reference. Additionally, if there exists an unknown or unstable gap between probe and sample, this relation can be such complex that a nonreferenced measurement of stress can be difficult. On the other hand, some global load estimate can enhance the results.

FLEXIBLE RISERS

Flexible risers are flexible pipes which are generally used to link subsea pipelines to floating offshore installations, such as FPSOs (floating production/storage/off-loading unit). In deep water oil and gas exploration flexible risers are used for oil and gas production, water and gas injection, and oil well control and monitoring [3]. Flexible risers are also used for oil and gas exportation to the shore or to a storage unit, such as FSOs (floating storage/off-loading unit).

A flexible pipe is made up of several different layers. The main components are leakproof thermoplastic barriers and corrosion-resistant steel wires. The helically wound steel wires give the structure its high-pressure resistance and excellent bending characteristics, thus providing flexibility and superior dynamic behavior. This modular construction, where the layers are independent but designed to interact with one another, means that each layer can be made fit-for-purpose and independently adjusted to best meet a specific field development requirement [4]. Figure 1 shows an example of flexible riser and Table 1 summarizes the function of each layer.

Table 1: Layers' functions

Layer	Function
Carcass	Prevent collapse
Internal pressure sheath	Internal fluid integrity
Interlocked pressure armor	Hoop stress resistance
Back-up pressure armor	Hoop stress resistance
Antiwear layer	Reduce friction between layers
Inner layer of tensile armor	Crosswound armor wires used for tensile stress resistance balanced with outer layer
Antiwear layer	Reduce friction between layers

Outer layer of tensile armor	Crosswound armor wires used for tensile stress resistance balanced with inner layer
Outer sheath	External fluid integrity

Pipa et al. EURASIP Journal on Advances in Signal Processing 2010 2010:176203 doi:10.1155/2010/176203

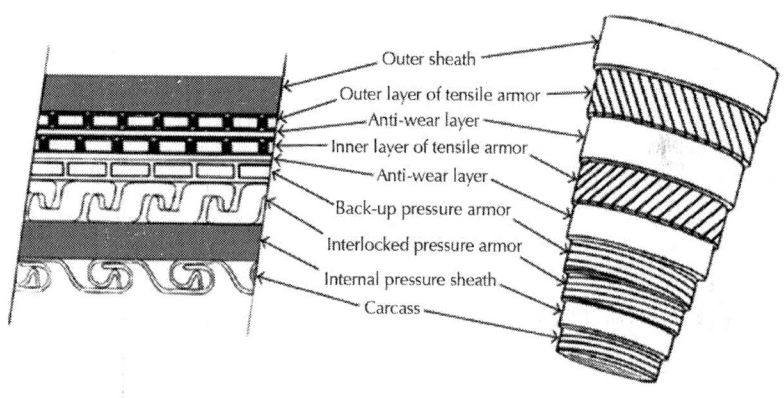

Figure 1: Unbonded flexible pipe.

This paper focuses on the integrity monitoring of the outer layer of tensile armor, which supports axial load and the riser weight. Its integrity is important to maintain a reliable connection between riser and floating unit (i.e., FPSO or platform).

Riser Failure Modes

Failure modes denote possible processes which cause the failure of a flexible pipe. In practice, a failure constitutes a loss of ability to transport product safely and effectively. This may be catastrophic (the pipe ruptures or breaks) or may constitute a minor, uncontrolled loss of pipe integrity or pipe blockage [5].

Table 2 describes the main failure modes of flexible pipes. The inspection and monitoring techniques suggested to detect and/or predict each failure mode are described in Section 2.2.

Table 2: Main failure modes of flexible pipes

No.	Failure mode	Description
1	Collapse	Collapse of carcass and/or pressure armor due to excessive tension, excessive external pressure or installation overloads
2	Burst	Rupture of tensile or pressure armors due to excess internal pressure
3	Tensile failure	Rupture of tensile armors due to excess tension
4	Compressive failure	Birdcaging of tensile armor wires
5	Overbending	Rupture or crack of external or internal sheaths
6	Torsional failure	Failure of tensile armor wires
7	Fatigue failure	Tensile armor wire fatigue
8	Erosion	Of internal carcass
9	Corrosion	Of internal carcass or tensile/pressure armor exposed to seawater or diffused product

Pipa et al. EURASIP Journal on Advances in Signal Processing 2010 2010:176203 doi:10.1155/2010/176203

Periodic inspections have detected a considerable incidence of damage in the top section of risers (i.e., end-fitting, Figure 2), which may affect their structural integrity and eventually induce different failure mechanisms. These include mostly external sheath damage, corrosion, and/or fatigue-induced damage to the tensile armors and torsional instability. These flaws are generally originated during installation or, more frequently, during operation due to contact with another riser or the platform structure [2, 6]. Figure 3 shows an example of a failure where rupture of tensile wires occurred inside the end-fitting.

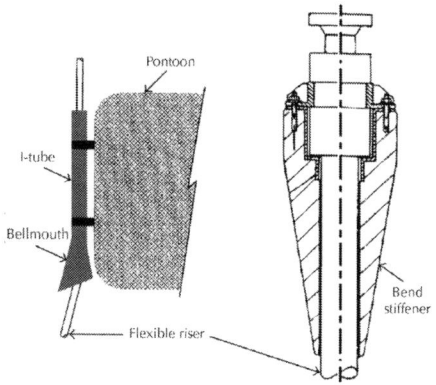

Figure 2: End-fitting.

Figure 3: End-fitting failure example.

Flexible Riser Inspection and Monitoring

The Recommended Practice for Flexible Pipe [7], also known as API 17B, from the American Petroleum Institute, recommends some inspection and monitoring methods for in-service flexible pipes. Table 3 lists the monitoring methods as well as the failure modes that are covered by each method. Visual inspection and periodic pressure testing have been, to date, the most common forms of in-service monitoring used for the demonstration of continued fitness for purpose.

Table 3: Flexible riser inspection and monitoring techniques suggested by API 17B and the failure modes (FM) covered by each method

Monitoring method	Description	FM covered
Visual inspection (internal and external)	Assessment of leakage or visible deformation or damage to pipe or outer sheath	1, 4, 5, 6, 8, 9
Pressure test	Pressure applied to pipe and decay measured as a function of time. Leakages or anomalies identified	1, 2, 5
Destructive analysis of removed samples	Prediction of the state of aging or degradation	8, 9
Load, deformation and environment monitoring	Measured parameters include wind, wave or current environment, vessel motions, product temperature, pressure and composition, and structural (or flexible pipe) loads and deformations	7
Nondestru-ctive testing of pipes in service	Radiography to establish the condition of steel tensile armor and pressure armor layers in service	2, 3, 4, 6, 7
Gaging operations	Gaging pigs to check for damage to the internal pipe profile	1, 8, 9
Spool Piece	To predict the state of aging or degradation of the internal pressure sheath	8, 9
Test pipe	Use of a flexible test pipe in series or in parallel with the flow which is periodically removed for destructive or nondestructive testing	8, 9

Annulus Monitoring	Measurement of annulus fluid (pH, chemical composition, volume). Prediction of degradation of the steel pressure armor or tensile armor layers or the aged condition of the internal pressure sheath or susceptibility of annulus environment to such degradation	7, 9

Pipa et al. EURASIP Journal on Advances in Signal Processing 2010 2010:176203 doi:10.1155/2010/176203

Several methods for managing the integrity of flexible pipes have been proposed in literature depending on the failure mode aimed [8–18]. As this work focuses on the detection of damage to tensile armor wires, the following survey of state-of-the-art methods will concentrate on techniques which directly or indirectly estimate the number of broken wires in armor layers.

Automated visual inspection has been employed by Petrobras as torsion monitoring. The method focuses on small angle deformation detection and on online data acquisition, in order to provide immediate identification for nonconformities. It consists of attaching a target on the riser and observing its behavior through a video camera, installed above the end fitting. The rupture of wires in the inner or outer layer can lead the riser to an unbalanced condition, thus generating torsion [2, 19]. However, if the number of broken wires is not significant, torsion might not occur and broken wires might not be detected.

Acoustic emission has been applied to detect the instant of rupture of armor wires. An acoustic emission scheme was developed in [20] based on laboratory tests and field experience. The idea is that when a tensile armor wire rupture takes place, a strong sound signal is generated. This great amplitude and energy sound wave can be distinguished, in relation to environmental noise, making acoustic emission a potential wire rupture detection technique. In [20], a procedure was designed to filter relevant acoustic events from spurious noisy emissions. The filtering scheme was applied to real data from a riser installed in the field. The riser was monitored for 11 months and then a dissection was performed. As no failure was found during the dissection, the filter parameters were adjusted to match the observed results. A drawback of this method is the need of continuous monitoring; that is, the rupture is not detected if the system is momentarily off. Also, other acoustic noises from the platform can cause false indications.

Fiber-optics Bragg grating (FBG) sensors technology has also been used to monitor flexible riser. In [21] two methodologies to monitor strain in flexible risers are developed. In the first approach, permanent FBG strain gages were installed on all wires of the armor layer. This approach allows identification of abrupt changes in the strain states of the wires, which may provide instantaneously detection of failure in one or more wires. The problem is that this method requires that the outer sheath be partially removed to access the wires. This is not always permitted in in-service risers.

In the second methodology, a thin steel collar instrumented with FBG strain gages was placed around the riser outer layer, measuring circumferential strains and changes in its diameter. Wire failures can be detected as they can cause variation in the external diameter of the polymeric jacket covering the riser. The disadvantage of this technique is that the number of broken wires needed to cause a detectable variation in the external diameter can be significant.

Another scheme using FBG strain gages was proposed in [1]. It is based on a retrofit clamp that monitors axial elongation and torsion of a flexible riser. The clamp is instrumented with FBG strain gages. As the previously presented methods, it suffers from sensibility. That is, one single broken wire is unlikely to be detected as its effects on external geometry are minimal.

In [22] a technology that integrates FBG sensors along grooves in the tensile wires during manufacturing of the pipe is described. Thus, strain and temperature can be monitored along several meters of the wires and ruptures are easily detected. Although new flexible pipes can be manufactured with this feature, the technology cannot be applied to existing pipes.

The electromagnetic tool MAPS-FR, on which the proposed method is based, is described in [3]. This equipment can estimate the stress on armor wires in a noninvasive manner. Additionally, it is sensitive to a single broken wire since it does not depend on geometrical deformations in the external sheath. Section 3.1 is devoted to describe MAPS-FR.

Basically, most techniques that directly estimate the number of broken wires are intrusive, whereas nonintrusive techniques are not precise. It will be shown that the proposed methodology, on the other hand, combines both advantages detecting a single wire break in a nonintrusive manner. Moreover, the proposed method produces graphical representation of stress distribution on wires which can be effectively used for break detection.

PROPOSED METHOD

This section describes the proposed method. In Section 3.1 the electromagnetic equipment used to measure internal tensile stress in the wires is briefly presented. The RLS filter technique is deduced in Section 3.2. Finally, the hybrid approach, which combines MAPS-FR signals with optical strain gage data through RLS filtering, is presented in Section 3.3.

MAPS-FR: Stress Measurement Technology

The tool used as nonintrusive stress gage for tensile armors is MAPS-FR. This equipment was developed by MAPS Technology in partnership with Petrobras. At the end of development process, Petrobras acquired the tool and has been developing its own signal processing algorithms, which are the main objective of this document. In the next lines, a brief presentation of MAPS-FR [23] tool is made. For a more complete description of MAPS-FR see [3].

In service the axial armor wires are subjected to tensile stress. In a failed wire, however, the applied tensile stress will be zero at the point of failure and will increase over some distance along the ligament from the break. The length over which the stress is increased depends on the amount of frictional load transfer to adjacent ligaments. If the length over which the stress reduction occurs is sufficiently long and the stress in the armor layer wires can be monitored, this would offer a method for detecting armor failure remotely from the actual failure location [3].

Most stress measurement techniques are not appropriate to monitor riser armor layers. Some, such as hole drilling, are clearly not satisfactory as they are not nondestructive and require access to the armor wires. Others, including neutron diffraction, and x-ray diffraction, are not suited to installed operation or, like ultrasonic methods, also need to be directly coupled to the material being measured [3].

Magnetic methods, on the other hand, do have the necessary attributes for an appropriate technology as intimate contact with the material being measured is not necessary [24, 25]. Stress measurement is possible as it is known that the magnetic properties of ferromagnetic materials are sensitive to internal stress. However, there are important issues to overcome as it is also known that mechanical hardness, grain size, texture, and other material properties also affect magnetic parameters. MAPS [23] stress measurement technology has been adapted to perform stress measurement in flexible pipes. This technique involves a number of low-frequency electromagnetic measurements, some of which monitor material variations, whilst others are mainly stress sensitive [3].

MAPS-FR Tool Description

The basic component of the current MAPS-FR equipment is the so-called probe, shown in Figure 4. Each probe contains an excitation coil, which generates the electromagnetic field that propagates through riser's wires, and three sensing coils, which read the response of a wire or group of wire to the excitation field. As previously mentioned, the value read by sensing coils depends on the stress that the wires are subjected to.

Figure 4: MAPS-FR probe.

Five probes are grouped together to form a ring, as shown in Figure 5. This assembly can now be mounted and fixed around the outer layer of the riser. As each probe has three sensing coils, a ring has fifteen sensing coils. The complete MAPS-FR equipment is composed by three rings, comprehending 45 sensing coils. Hence, the current MAPS-FR set permits monitoring of approximately 45 wires on the external armor layer, although this can be altered to suit requirements.

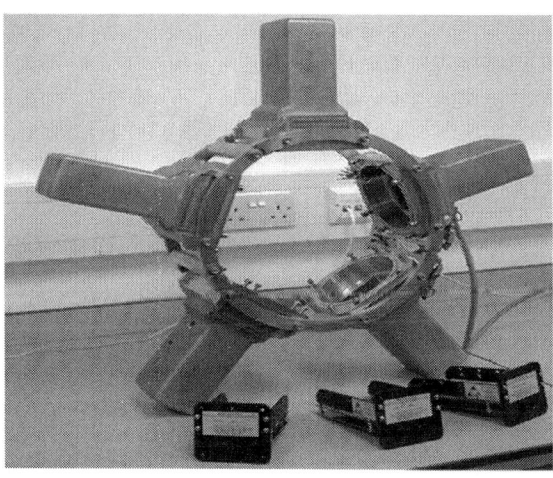

Figure 5: MAPS-FR ring.

MAPS-FR Data

The goal achieved by current MAPS-FR technology is to compare tensile stress present in armor wires. Nonetheless, the interpretation of raw data requires an analysis by MAPS-FR experts. As a result, an indication of a possible wire rupture is signalized including its circumferential localization.

During the development of MAPS-FR system, Petrobras and MAPS Technology jointly performed several controlled laboratory tests. In these tests, specific wires were induced to failure by the introduction of notches on their surfaces. Blind tests were also performed, where only the Petrobras team was acquainted with damage wires and the MAPS Technology team had to give indication of broken wires based only on MAPS-FR's signals. In the final blind test, MAPS-FR experts correctly indicated 100% of wire breaks with 1 false positive indication over 9 correct indications.

The raw MAPS-FR signals exhibit a slow time drift probably due to accommodations of riser's internal layers during a load variation. This drift must be carefully considered in order not to be misinterpreted as a wire break. Automatic break detection algorithms have to compensate these phenomena avoiding false calls. In a nonreferenced monitoring, that is, when MAPS-FR operates without a global riser load estimate, one cannot say whether this drift is actually a load change or only the drift behavior.

When continuously operating in a off-shore environment, MAPS-FR can generate huge amounts of data, yielding the human-based interpretation arduous and unfeasible. An automatic approach is essential as a preliminary analysis, signalizing only important events to be reviewed by experts. The method proposed in this document is the first step towards an automatic wire break detection system.

RLS Adaptive Filtering

Filters are a particular important class of linear time-invariant systems [26]. Strictly speaking, the term frequency-selective filter suggests a system that passes certain frequency components and totally rejects all others, but in a broader context any system that modifies certain frequencies relative to others is also called a filter [27]. Another meaningful definition is that filter is a device that maps its input signal to another output signal facilitating the extraction of the desired information contained in the input signal [28]. The latter definition is particularly interesting in the context of this document.

Adaptive filters are, in turn, time-varying systems which adapt their parameters to a more suitable condition or operation point in order to achieve a specified behavior. In other words, the filter coefficients are changed so as an input signal is transformed in an output signal which is as equal as possible to a desired signal.

RLS adaptive filter class aims at the minimization of the sum of the squares of the difference between the desired signal and the filter output signal. When new samples of incoming signals are received at every iteration, the solution for the least-squares problem can be computed in recursive form resulting in the recursive least-squares (RLSs) algorithms [28].

Let x(n) be the input signal, let y(n) be the output signal, and let d(n) be the desired signal, with n representing the time. That is, d(0) is the value of desired signal at time 0. The input vector is formed by the last N+1 values of the input signal and is given by

$$\mathbf{x}(n) = [x(n) \ x(n-1) \ \cdots \ x(n-N)]^T. \tag{1}$$

The filter, which transforms the input signal x(n) into the output y(n) is given by

$$\mathbf{w}(n) = [w_0(n) \ w_1(n) \ \cdots \ w_N(n)]^T, \tag{2}$$

where n is the filter order. Note that due to its adaptive nature, the filter coefficients w(n) are time-varying, denoted by letter n. The output signal at any instant n can be obtained by

$$y(n) = \mathbf{x}^T(n)\mathbf{w}(n-1). \tag{3}$$

The prediction error is at instant n given by

$$e(n) = d(n) - y(n). \tag{4}$$

Figure 6 depicts the general scheme of an adaptive filter. An adaptive algorithm adjusts the main filter coefficients based on some metric applied to the output error e(n). In general, the adaptive algorithm will choose the main filter parameters so that the output error e(n) is minimized.

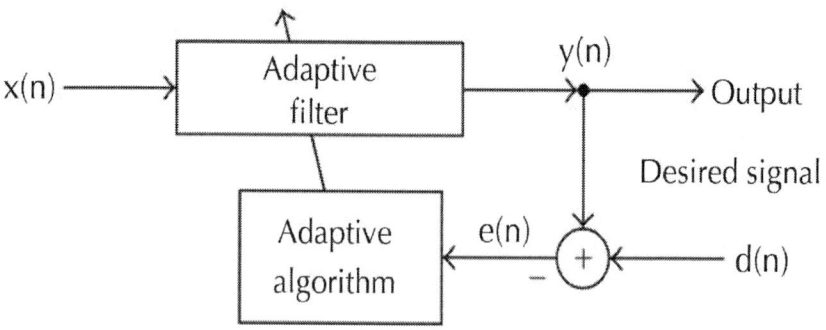

Figure 6: General adaptive filter configuration.

The goal of RLS methods is to minimize not only the last error but also the sum of all past output errors. Thus, the objective function is given by

$$\xi(n) = \sum_{i=0}^{n} \lambda^{n-i} \varepsilon^2(i) = \sum_{i=0}^{n} \lambda^{n-i} \left[d(i) - \mathbf{x}^T(i)\mathbf{w}(n) \right]^2, \tag{5}$$

where $0 < 1$ is an exponential weighting factor also referred to as forgetting factor. The forgetting factor permits to put more significance and weight on recent output errors than distant past errors. The lesser the forgetting factor is, the less important are old output errors to the coefficient updating.

By differentiating (n) with respect to w(n) in (5) and performing some algebraic manipulations, the final algorithm, shown in Algorithm 1, can be deduced.

Algorithm 1: Complete RLS algorithm.

Initialization

$S(-1) = \delta I$

$w(-1) = [0 \ 0 \ \cdots \ 0]^T$

Do for $n \geq 0$:

$e(n) = d(n) - x^T(n)w(n-1)$

$\psi(n) = S(n-1)x(n)$

$S(n) = \frac{1}{\lambda}\left[S(n-1) - \frac{\psi(n)\psi^T(n)}{\lambda + \psi^T(n)x(n)}\right]$

$w(n) = w(n-1) + e(n)S(n)x(n)$

$y(n) = x^T(n)w(n)$

$\varepsilon(n) = d(n) - y(n)$

The RLSs are known to pursue fast convergence and have excellent performance when working in time-varying environments. See [28, 29] for more information on adaptive filters and RLS algorithms.

Hybrid Approach

The current MAPS-FR tool uses 4 excitation frequencies during the acquisition yielding 4 signals per sensing coil. Each frequency shows a different sensitivity to wire stress depending on wire size, wire depth, and so forth. It will be shown that a proper combination of the 4 signals per sensing coil gives a better estimate of the stress than the one obtained by considering each signal independently.

The idea of the hybrid approach is to find a set of linear systems which map each of MAPS-FR sensing coil's signals into realistic load values. These linear systems are continuously recalculated at every iteration to compensate the slow time drift exhibited by MAPS-FR signals. Although magnetic properties of metals vary nonlinearly with mechanical load, linear systems can be used to do this mapping if some adaptation is permitted. That is, the correspondence holds (i.e., mapping becomes linear) in a small region surrounding a given operation point. Once the operation point changes, the adaptive filter recalculates its coefficients. The new filter coefficients are valid within this new region.

The hybrid approach needs an estimate of riser global load to be used as the desired signal $d(n)$. Indeed, if all wires are unbroken, the riser global load is approximately equally divided to each wire and it can be used as an estimate of stress in each wire. Since only differences between wire stresses are important to detect a broken wire, the riser global load is taken as desired signal for each wire (Section 3.2).

Problem Statement

MAPS-FR signals are arranged in a matrix form as

$$\mathbf{X}(n) = \begin{bmatrix} x_1^{(1)}(n) & x_2^{(1)}(n) & \cdots & x_{45}^{(1)}(n) \\ x_1^{(1)}(n-1) & x_2^{(1)}(n-1) & \cdots & x_{45}^{(1)}(n-1) \\ \vdots & \vdots & \ddots & \vdots \\ x_1^{(1)}(n-N) & x_2^{(1)}(n-N) & \cdots & x_{45}^{(1)}(n-N) \\ x_1^{(2)}(n) & x_2^{(2)}(n) & \cdots & x_{45}^{(2)}(n) \\ x_1^{(2)}(n-1) & x_2^{(2)}(n-1) & \cdots & x_{45}^{(2)}(n-1) \\ \vdots & \vdots & \ddots & \vdots \\ x_1^{(2)}(n-N) & x_2^{(2)}(n-N) & \cdots & x_{45}^{(2)}(n-N) \\ x_1^{(3)}(n) & x_2^{(3)}(n) & \cdots & x_{45}^{(3)}(n) \\ x_1^{(3)}(n-1) & x_2^{(3)}(n-1) & \cdots & x_{45}^{(3)}(n-1) \\ \vdots & \vdots & \ddots & \vdots \\ x_1^{(3)}(n-N) & x_2^{(3)}(n-N) & \cdots & x_{45}^{(3)}(n-N) \\ x_1^{(4)}(n) & x_2^{(4)}(n) & \cdots & x_{45}^{(4)}(n) \\ x_1^{(4)}(n-1) & x_2^{(4)}(n-1) & \cdots & x_{45}^{(4)}(n-1) \\ \vdots & \vdots & \ddots & \vdots \\ x_1^{(4)}(n-N) & x_2^{(4)}(n-N) & \cdots & x_{45}^{(4)}(n-N) \\ 1 & 1 & \cdots & 1 \end{bmatrix},$$

(6)

where $x^{(p)}_q(n)$ is the value of qth MAPS-FR sensing coil at frequency p at instant n. The matrix $X(n)$ is of size $M \times K$, where $M = 4N + 1$ are the N past values of MAPS-FR signal for each of 4 frequencies, plus a constant for bias correction, and K = 45 are the 45 sensing coils. The filter coefficients are given by

$$W(n) = \begin{bmatrix} w_{1,1}(n) & w_{1,2}(n) & \cdots & w_{1,45}(n) \\ w_{2,1}(n) & w_{2,2}(n) & \cdots & w_{2,45}(n) \\ \vdots & \vdots & \ddots & \vdots \\ w_{M,1}(n) & w_{M,2}(n) & \cdots & w_{M,45}(n) \end{bmatrix}.$$

(7)

A wire's stress estimate is given by

$$y_k(n) = \mathbf{x}_k^T(n)\mathbf{w}_k(n-1),$$

(8)

where $y_k(n)$ is the estimate of stress in kth wire at instant n, $x_k(n)$ is the kth column of matrix X(n), and $w_k(n-1)$ is the k column of matrix W(n − 1). Arranging yk(n) in a vector form, one can write y(n) = [y_1(n) y_2(n) ⋯ y_{45}(n)]^T. As the only reference available for the riser load is d(n), it can be written as a vector d(n) = [d(n) d(n) ⋯ d(n)]^T = d(n)[1 1 ⋯ 1]^T.

The estimate error can, thus, be written in vector shape as

$$\mathbf{e}(n) = \mathbf{y}(n) - \mathbf{d}(n).$$

(9)

Figure 7 shows the block diagram of the hybrid adaptive filter. A summary of variables is given in Table 4 and the complete algorithm is shown in Algorithm 2.

Table 4: Hybrid algorithm variables

Variable	Meaning
d(n) = [d(n) d(n) ⋯ d(n)]^T	Desired signal vector: global riser load estimate obtained from optical strain gages.
X(n) = [x1(n) ⋯ x45(n)]	MAPS-FR signals matrix organized as in (6).
y(n) = [y_1(n) y_2(n) ⋯ y_{45}(n)]^T	Output vector which will be plotted for wire break detection.

W(n) = [w1(n) ⋯ w45(n)]	Adaptive filter matrix coefficients as in (7) that are updated at every iteration.

Pipa et al. EURASIP Journal on Advances in Signal Processing 2010 2010:176203 doi:10.1155/2010/176203

Algorithm 2: Complete Hybrid algorithm.

Initialization

Do for $0 \leq k \leq 45$

$$w_k(-1) = [1\ 1\ \cdots\ 1]^T$$

Do for $n \geq 0$

Do for $0 \leq k \leq 45$

$$y_k(n) = x_k^T(n) w_k(n-1)$$

Update $w_k(n)$ for each k as in Algorithm 1

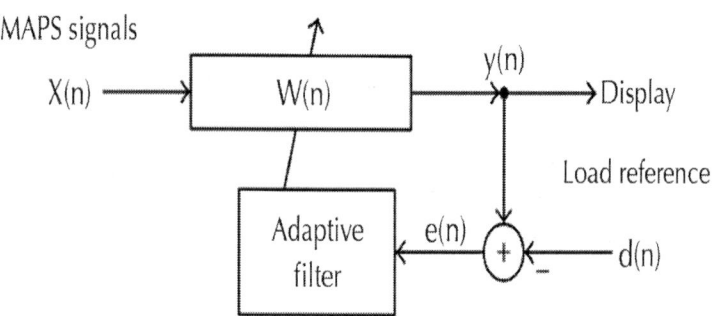

Figure 7: Hybrid adaptive filter.

Notice that $y_k(n)$ is calculated with the filter coefficients $w_k(n-1)$ obtained in the previously iteration. Consequently if a wire break occurs between time n − 1 and n at, say, wire k = k_0, the signal y_{k0} (n) tends to abruptly diverges from other $y_k(n)$'s at instant n, indicating the rupture. As time passes, w_{k0} (n) are recalculated for the new condition and the divergence between k_0 and other k's vanishes. The vanishing behavior is explained next: suppose that there are as many sensing coils as wires in the external armor layer. Even if each sensing coil is located exactly above each wire, respectively, due to the gap between sensing coil and wire (i.e., external polymer layer thickness), magnetic field from adjacent wires leaks laterally and affects the measurements of each other. Therefore, there exists some portion of magnetic field surrounding a broken wire that contains signals from unbroken wires. The adaptive filter parameters $wk_0(n)$ are recalculated based on this portion of signal which is coherent to the loading $d(n)$. Although this seems to be a problem, detectability at exact instant of rupture is unaffected, as will be shown in Section 4.

The units of stress and magnetic fields are irrelevant in this context. The whole system works by comparison; that is, the goal is to determine whether there is an inactive wire among active ones. Nevertheless, it is possible to establish a calibration procedure which would give rise to consistent measurements, though it is out of the scope of this document.

RESULTS

A trial has been carried out to evaluate the performance of the proposed method. The trial took place at the riser fatigue test rig of Physical Metallurgy Laboratory (LAMEF) in Porto Alegre, Brazil. The facility allows the application of static and dynamic tensile loads exceeding 220 tons. The test sample was a section of 6" nominal bore new flexible production riser, rated for 3000 psi of approximately 5 m length. One end of the riser was fixed and subjected to axial load (nominally connector A), whereas the other was free to rotate during loading (nominally connector B). The loading was cyclic and sinusoidal, from 160 to 220 ton and at a frequency of 0.0167 Hz.

In the tests, the riser loadings were chosen to simulate as accurately as possible the field conditions, namely, approximately 30% to 50% of yield point. However, other field conditions such as internal pressure, arbitrary load instead of cycling load, and riser orientation (vertical instead of horizontal) were not considered. The influence of these circumstances on the results is intended to be studied in future tests.

Two windows were opened on the external sheath to access the wires. The first window was near to connector B and had a circumferential shape, giving access to all wires. This window (right side of Figure 9) was used to cut the wires during loading, simulating a real rupture. The second window was close to connector A and was used to instrument all wires with optical strain gages. The signals from these strain gages were used as references (i.e., real stress of wires). The global riser load needed for the hybrid processing was estimated averaging all strain gages signals.

The MAPS-FR was installed in the middle of the sample. This configuration was chosen to ensure that the strain gages measured similar tension values to those sensed by the MAPS-FR.

Figure 8 shows a comparison between pure MAPS-FR signals and the signal obtained by the proposed method. Pure MAPS-FR signals do not represent the load well. However, when properly combined, it is possible to obtain a signal which is similar to the load that the riser is subjected to.

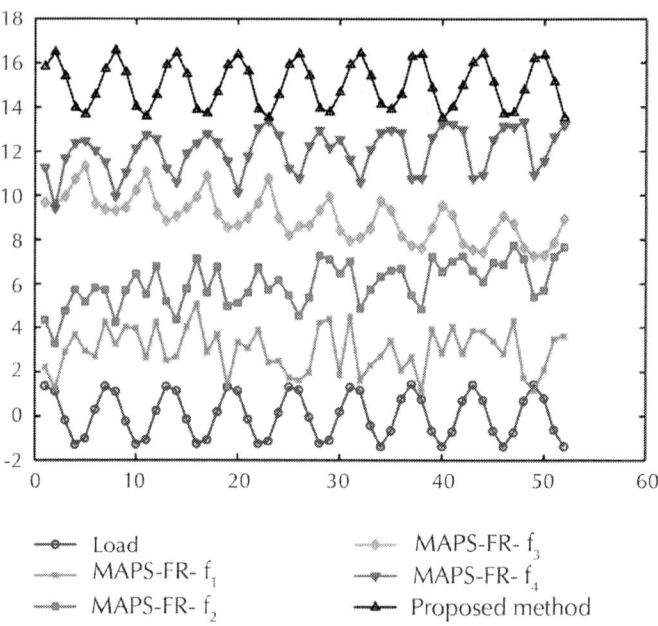

Load
MAPS-FR- f_1
MAPS-FR- f_2
MAPS-FR- f_3
MAPS-FR- f_4
Proposed method

Figure 8: Pure MAPS-FR signals aims to estimate the riser load. However, better results are achieved when the four signals are combined by the proposed method.

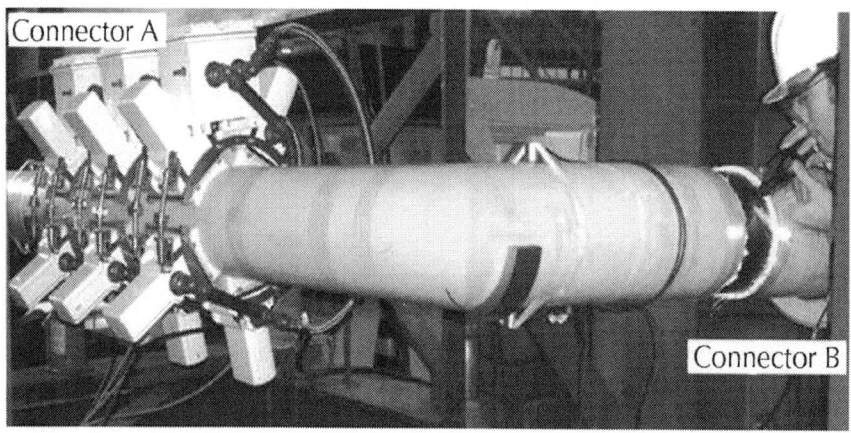

Figure 9: Trial.

Table 5 summarizes the mean-square error between pure MAPS-FR signals and the loading. The phase difference between the signals was corrected previously to the error computation. The proposed method's signal is the one that best represents the riser load.

Table 5: Comparison between pure MAPS-FR signals and proposed method's signal and load signal

Signal	MSE
MAPS-FR frequency 1	1.1164
MAPS-FR frequency 2	0.9781
MAPS-FR frequency 3	0.7956
MAPS-FR frequency 4	0.5291
Proposed method	0.2676

Pipa et al. EURASIP Journal on Advances in Signal Processing 2010 2010:176203, doi:10.1155/2010/176203

Figure 9 illustrates the setup and Table 6 summarizes the trial resources and details.

Table 6: Trial details

Laboratory	Laboratory LAMEF at UFRGS University
Local	Porto Alegre-RS-Brazil
Date	2009/06/24 and 25
Sample	5 meters long 6"riser sample
Total external diameter	250.7 mm
External sheath thickness	7 mm
Number of external sheaths	2
MAPS-FR mounted diameter	(1 sheath removed) 236.7 mm
External armor layer	37 flat wires: 15 mm × 5 mm at 30 degrees
Reference instrumentation	Optical strain gages on every wire
Global riser load estimate d(n)	Average of optical strain gages
Loading	Sinusoidal, 160 to 220 ton, frequency 0.0167 Hz
Magnetic stress measuring tool	MAPS-FR
Signal processing algorithm	Hybrid MAPS-FR/Optical strain gage

Pipa et al. EURASIP Journal on Advances in Signal Processing 2010 2010:176203, doi:10.1155/2010/176203

The main events of the trial are listed in Table 7. As already mentioned, the wire breaks were produced cutting them at the first window, close to connector B. In all events, the optical strain gages almost instantaneously detected the respective wire rupture. In some cases, the hybrid algorithm indicated the rupture seconds before optical strain gages.

Table 7: Important events

Date	Time	Time stamp	Event
2009/06/24	15 h 15	250	Wire 37 break
	16 h 30	439	Wire 35 break
	17 h 12	589	Wire 30 break
	17 h 38	690	Wire 6 break

2009/06/25	13 h 34	330	Wire 5 break
	14 h 36	523	Wire 17 break
	16 h 14	823	Wire 7 break
	17 h 52	1160	Wire 13 break
	18 h 21	1258	Wire 27 break

Pipa et al. EURASIP Journal on Advances in Signal Processing 2010 2010:176203, doi:10.1155/2010/176203

Figure 10 illustrates a series of cyclic loads. Each of the following figures is composed of two graphs: the optical strain gages' signals are shown on the left, and processed MAPS-FR signals are shown on the right. The abscissae represent time stamp, whereas the ordinates represent wire number (a) or sensing coil number (b). Given a wire number and a time instant, the correspondent color indicates the stress level. In both graphs, there is a scale bar on the right which relates the color to a statistically normalized stress parameter.

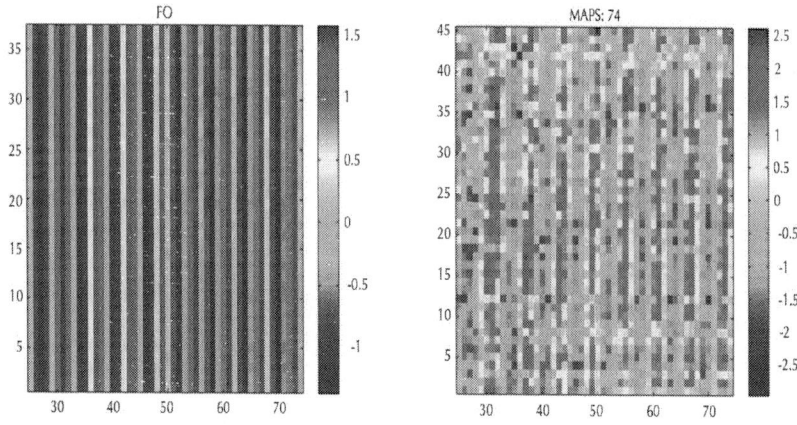

Figure 10: Normal cyclic loading.

Figures 11 to 19 show the rupture moment of several wires. It is clear on the left graphs which wire was cut. The color difference indicates that the damaged wire diverges from the unbroken ones; that is, cut wires tend to loosen. Although there is some noise on the right graphs, it is possible to determine the rupture instant in most cases. Moreover, the scale range of processed MAPS-FR signals remains between -3 and +3 during normal operation. When a rupture occurs, its limits reach from ±3.5 to ±20. This could be used as a criterion for automatic detection.

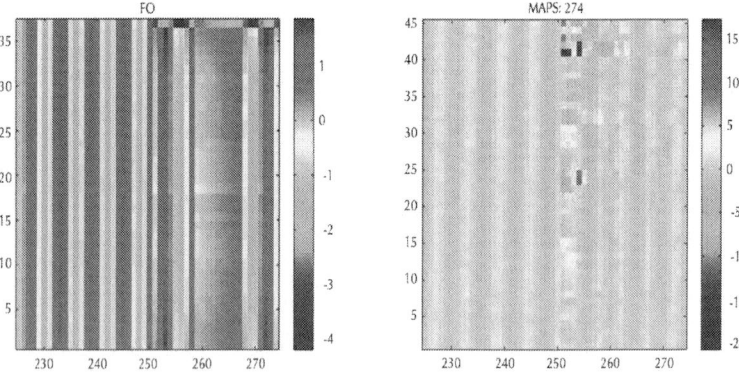

Figure 11: Wire 37 break.

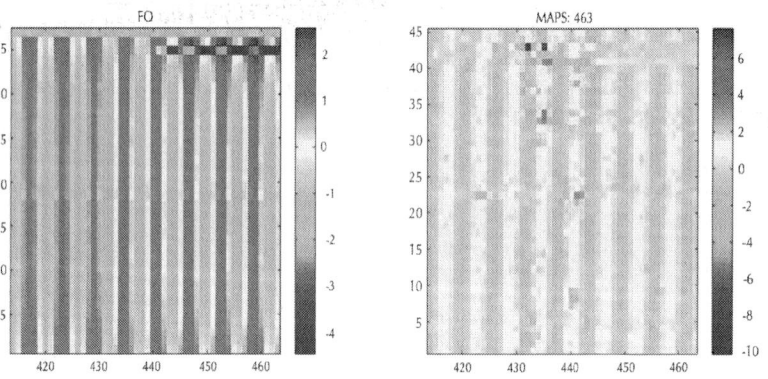

Figure 12: Wire 35 break.

Flexible Riser Monitoring using Hybrid Magnetic/Optical Strain.... 151

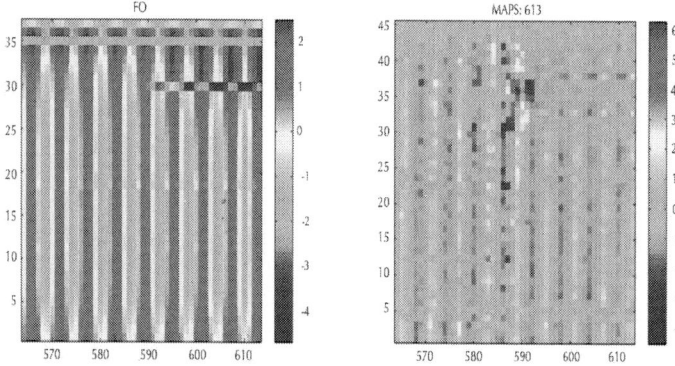

Figure 13: Wire 30 break.

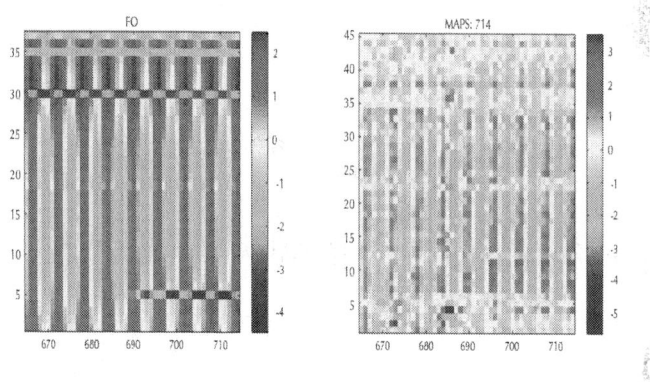

Figure 14: Wire 6 break.

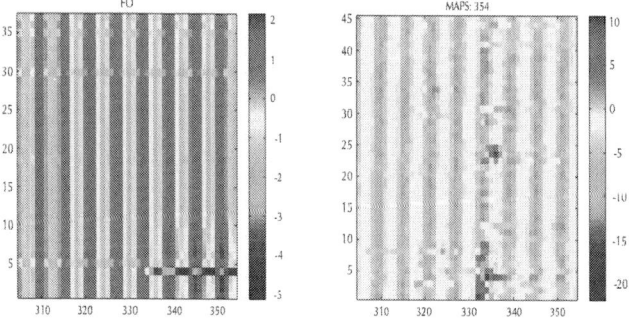

Figure 15: Wire 5 break.

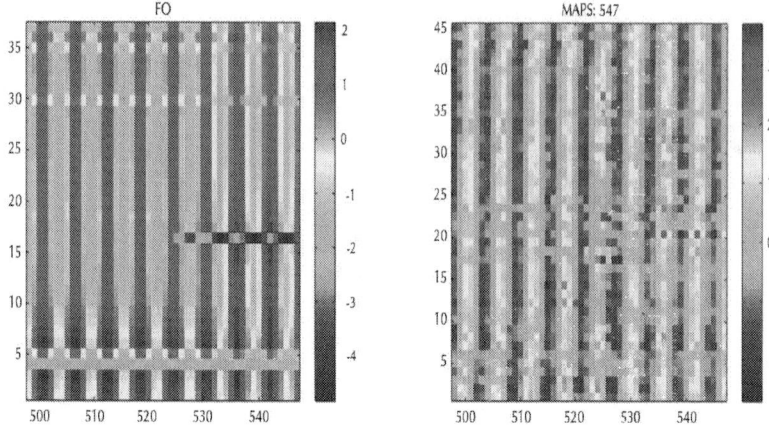

Figure 16: Wire 17 break.

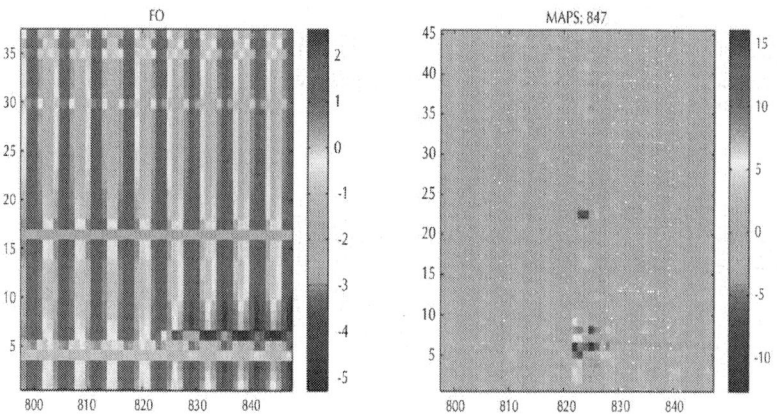

Figure 17: Wire 7 break.

Flexible Riser Monitoring using Hybrid Magnetic/Optical Strain....   153

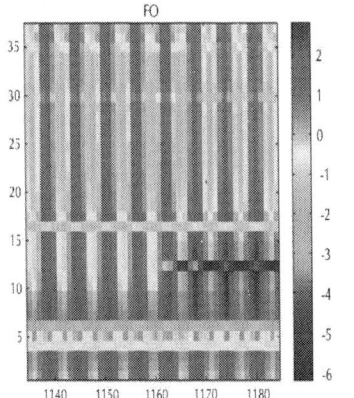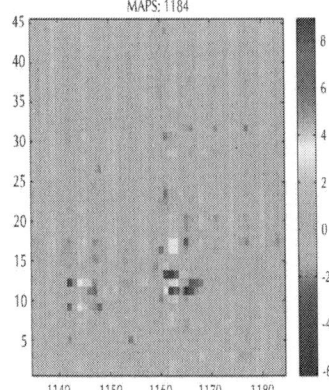

Figure 18: Wire 13 break.

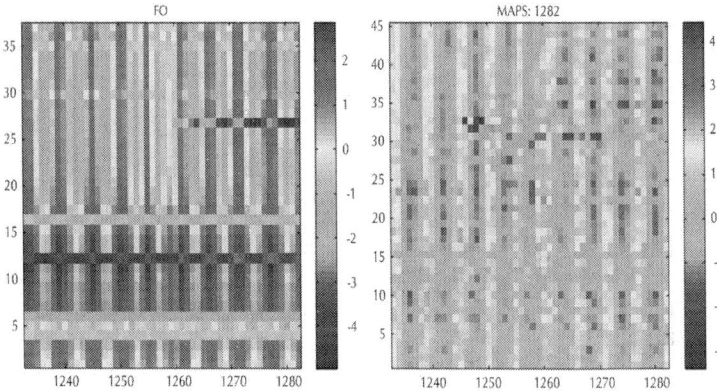

Figure 19: Wire 27 break.

CONCLUSIONS

Flexible risers are multilayered pipes used in oil and gas industry. Their complex geometry imposes difficulties (i.e., unknown wire arrangement) when assessing stress in internal layers through the outer polymeric sheath. Unknown wire arrangement and in-service wire reaccommodation introduce uncertainties while measuring internal stress.

This article proposes a new estimation method of internal stress distribution by combining electromagnetic measurements with optical strain gage data. Electromagnetic measurements are converted into load values through adaptive filters. Optical strain gage signal is used as an estimate of riser global load. This signal is used as desired signal in the adaptive context. In other words, it is assumed that in normal conditions riser global load is equally divided between wires.

A set of adaptive linear filters is calculated so as each of the MAPS-FR signals is converted into load. The filters' inputs are electromagnetic signals, and the filters' outputs are load estimates of correspondent wires. When a wire rupture occurs, the filter produces indications that the load changed, and the break can be detected.

The main advantage of the proposed technique is that it does not need the external sheath to be removed; that is, it is a nonintrusive technique. Yet, it can detect single-wire ruptures. The riser global load estimate, required by the proposed method, can be obtained through strain gages installed on the external sheath or by the FBG strain gage collar mentioned in Section 2.2 and described in [21]. The collar can detect and estimate the riser global load in a nonintrusive manner.

The presented results showed that the proposed technique produces graphical representations on which visual detection of wire breaks can be effectively performed in most cases. The proposed method can be straightforwardly extended to automatically detect wire ruptures. A simple fixed threshold or statistically variable threshold could be employed for this purpose.

ACKNOWLEDGMENTS

The authors would like to acknowledge Petrobras Research and Development Center (CENPES) for its incentive and financial support. The authors would like to acknowledge also MAPS Technology and MAPS team for the partnership, collaboration, corrections, and experience exchange during the production of this work.

REFERENCES

1. H Corrignan, RT Ramos, RJ Smith, S Kimminau, L El Hares, New monitoring technology for detection of flexible armor wire failure. Proceedings of the Offshore Technology Conference (OTC '09), 2009
2. MG Marinho, CS Camerini, SR Morikawa, DR Pipa, GP Pires, JM Santos, New techniques for integrity management of flexible riser end-fitting connection. Proceedings of the 27th International Conference on Offshore Mechanics and Arctic Engineering, June 2008, Estoril, Portugal
3. JC McCarthy, DJ Buttle, Non-invasive magnetic inspection of flexible riser. Proceedings of the Offshore Technology Conference (OTC '09), May 2009, Houston, Tex, USA
4. Technip, Flexible pipe brochure (2008, http://www), . technip.com/pdf/Flexible_Pipe.pdf webcite
5. Health and S. Executive, Guidelines for integrity monitoring of unbonded flexible pipe (Health and Safety Executive, 1998)
6. MG Marinho, JM dos Santos, RDO Carneval, Integrity assessment and repair techniques of flexible risers. Proceedings of the 25th International Conference on Offshore Mechanics and Arctic Engineering (OMAE '06), June 2006, Hamburg, Germany
7. A. P. Institute, API RP 17B—Recommended Practice for Flexible Pipe (API Publishing Services, Washington, DC, USA, 2002)
8. A Berg, NJ Rishøj-Nielsen, Integrity monitoring of flexible risers by optical fibres. Proceedings of the 21st International Conference on Offshore Mechanics and Arctic Engineering (OMAE '02), 2002 **3**, 47–52
9. LA Mesquita, JM Santos, P Loureiro, AL Carvalho, Monitoramento das válvulas de despressurização de gás percolado no espaço anular de risers de produção e exportação de óleo e gás. Proceedings of the Rio Pipeline Conference and Exposition, 2005
10. A Felix-Henry, P Lembeye, Flexible pipes in-service monitoring. Proceedings of the 23rd International Conference on Offshore Mechanics and Arctic Engineering (OMAE '04), 2004 **3**, 149–154
11. J Marsh, P Duncan, I MacLeod, Offshore pipeline and riser integrity—the big issues. Proceedings of the Offshore Technology Conference (OTC '09), 2009

12. C Saunders, T O'Sullivan, Integrity management and life extension of flexible pipe. Proceedings of the Offshore Technology Conference (OTC '07), 2007
13. JW Picksley, K Kavanagh, S Garnham, D Turner, Managing the integrity of flexible pipe field systems: industry guidelines and their application. Proceedings of the Annual Offshore Technology Conference, 2002, 609–618
14. J Picksley, State of the art flexible riser integrity issues: study report (MCS International, 2001)
15. N Weppenaar, A Kosterev, L Dong, D Tomazy, F Tittel, Fiberoptic gas monitoring of flexible risers. Proceedings of the Offshore Technology Conference (OTC '09), 2009
16. R Roberts, S Garnham, B D'All, Fatigue monitoring of flexible risers using novel shape-sensing technology. Proceedings of the Offshore Technology Conference (OTC '07), 2007
17. R Thethi, H Howells, S Natarajan, C Bridge, A fatigue monitoring strategy and implementation on a deepwater top tensioned riser. Proceedings of the Offshore Technology Conference (OTC '05), 2005
18. E Binet, P Tuset, S Mjøen, Monitoring of offshore pipes. Proceedings of the Offshore Technology Conference (OTC '03), 2003
19. MG Marinho, CS Camerini, JM Santos, GP Pires, Surface monitoring techniques for a continuous flexible riser integrity assessment. Proceedings of the Offshore Technology Conference (OTC '07), 2007, Houston, Tex, USA
20. SD Soares, CS Camerini, JMC de Santos, Development of flexible risers monitoring methodology using acoustic emission technology. Proceedings of the Offshore Technology Conference (OTC '09), 2009
21. SRK Morikawa, CS Camerini, DR Pipa, JMC Santos, GP Pires, AMB Braga, RWA Llerena, AS Ribeiro, Monitoring of flexible oil lines using FBG sensors. Proceedings of the 19th International Conference on Optical Fibre Sensors, April 2008, Proceedings of SPIE 7004, 70 046F-1–70 046F-4
22. M Andersen, A Berg, S Saevik, Development of an optical monitoring system for flexible risers. Proceedings of the Offshore Technology Conference (OTC '01), 2001

23. http://www.maps-technology.com/
24. B DJ, Emerging technologies for in-situ stress surveys. Proceedings of the 6th International Conference on Residual Stresses (ICRS '00), 2000
25. B DJ, S CB, Residual stresses: measurement using magnetoelastic effects. The Encyclopaedia of Materials: Science and Technology (2001)
26. PSR Diniz, EAB da Silva, SL Netto, Digital Signal Processing: System Analysis and Design(Cambridge University Press, Cambridge, UK, 2002)
27. JBAV Oppenheim, RW Schafer, Discrete-Time Signal Processing, 2nd edn. (Prentice-Hall, Upper Saddle River, NJ, USA, 1997)
28. PSR Diniz, Adaptive Filtering: Algorithms and Practical Implementations, 3rd edn. (Springer, Boston, Mass, USA, 2008)
29. S Haykin, Adaptive Fiter Theory, 3rd edn. (Prentice-Hall, Upper Saddle River, NJ, USA, 1996)

Chapter 6

Improve the Government Strategic Petroleum Reserves

Xiucheng Dong, Zhongbing Zhou, and Hui Li

China Oil & Gas Centre, Beijing, China

ABSTRACT

The potentiality that the current government strategic petroleum reserves (GSPRs) can be improved by the pre-allocation of GSPR drawing rights has been neglected. This paper proposes to pre-allocate the GSPR drawing rights, and proves that by doing this the efficiency of GSPR and the society's incentive to finance GSPR can be improved.

Particularly, the example demonstrates that the incentive improvement can be very significant. Since it takes huge expenditure on GSPR and it is very important to gain support from the consumers by improving GSPR, the proposal is quite worth considering.

INTRODUCTION

To prepare for oil supply interruption in advance, strategic petroleum reserves (SPRs) [1] across the countries with high dependency on imported oil have been built. Almost 1/3 of SPRs are government owned stocks [2], or what we call government strategic petroleum reserves (GSPRs). GSPR is vulnerable to criticisms and the society may have weak incentives to support it, because it is totally financed by public expenditure. The institution CATO published an analysis in 2005, arguing that the US SPR programs were inefficient since they have cost the US citizens too much but only generated a little benefit [3]. The criticisms on GSPRs motivated us to come up with a proposal for improving the GSPR. Based on strong assumptions, the proposal reserves a great possibility to improve the current GSPRs.

The assumptions and the proposal will be elaborated in Section 2. Section 3 shows that how the proposal is supposed to improve the current GSPR in terms of efficiency, and Section 4 proves that the proposal can increase the society's incentive to support GSPRs. The significance and limitedness of the proposal will be discussed in Section 5. Section 6 concludes the paper.

THE ASSUMPTIONS AND THE PROPOSAL

The Assumptions

SPRs generate two kinds of good. First, SPRs can keep oil price from soaring at supply interruption; second, SPR drawing rights enable the owners to be more competitive for the released SPR oil and thus give them advantages at supply interruption. The first good benefits all oil consumers and excludes none from enjoying it at bearable cost, so for it we have the name "public good", which is typically a kind of Samuelson's pure public goods [4]. The second good can easily exclude any one by price biding, so for it we have the name "private good".

Two government types are defined as $G_0(I,M,U_0,C,t,r)$ and $G_0(I,M,U_0+U_1,C,t,r)$ Where I is the number of the oil consumers, M is the size of the GSPR the government need build, U_0 is the aggregate utility function of GSPR as pure public good while U_1 is the aggregate utility function of the pre allocated GSPR drawing right, C is the GSPR's total cost, t is the anticipated duration between the time 0 and the next supply interruption, and r is the real rate incorporating factors of the oil price's long run trend. The difference between the two governments is that under G_0 GSPR is purely financed by public expenditure, while under G_1 GSPR is financed not only by public expenditure but also by the private bids for the GSPR drawing rights before the construction of GSPR. For the governments, assumptions 1-3 are arranged.

Assumption 1: Both G_0 and G_1 know the function C(M), which is concave and continuously increasing on M.

Assumption 2: G_0 knows U_0, and G_1 knows both U_0 and U_1.

Assumption 3: As net oil importer, both G_0 and G_1 maximizes the aggregate consumer surplus, which means $\partial U_0/\partial M = \partial C/\partial M$ must hold for G_0, and $\partial U_0/\partial M + \partial U_1/\partial M = \partial C/\partial M$ must hold for G_1.

Consumers are willing to pay for GSPR because they have expected an oil supply interruption which may endanger their welfare and this danger may cost them more than GSPR does. As price takers under the current oil market condition, oil consumers know well that only collective actions can deter the soaring oil price. Economies of scale provide a main rationale for considering public infrastructure provision [5]. GSPR projects have extraordinary economies of scale, and the total GSPR cost would be unacceptably high if construction actions have been taken separately rather than collectively.

Let $u_i(M)$ stand for consumer i's utility function of GSPR as public good, and $U_{id}(\lambda_i)$ stand for consumer i's utility of his/her GSPR drawing right λ_i, C_i stand for consumer i's payment for GSPR, B_i stand for i's proportional share of the total cost pie of the GSPR, and $\sum \lambda_i = M$ For the consumers, assumptions 4-6 are arranged.

Assumption 4: u_i is concave and continuously increasing on M; u_{id} is concave and continuously increasing on λ_i; $\sum u_i = U_0$, and $\sum u_{id} = U$.

Assumption 5: $\partial u_i / \partial M + \partial u_i / \partial \lambda_i = \partial C_i$. $C_i = B_i c(M)_i$, where $0 \leq B_i \leq 1$ and $\sum B_i = 1$.

Assumption 6: There exists certain $\underline{M}$ so that in $[0, \underline{M}]$, $C \leq U_0 + U_1$ while in $(\underline{M}, +\infty)$ $C > U_0 + U_1$.

It can be inferred that $U = \sum (u_i + u_{id})$ is also continuously increasing on M. However, there is free-riding problem in the provision of public good and it's rather difficult to overcome this problem by decentralized mechanisms, so for simplicity we assume that under the government' authority, the society's propensity to free-ride has been wiped out. Under this circumstance, the government can optimally decide the size of the GSPR as public good and the distribution of the corresponding cost. The rationale of the last sentence of assumption 6 is, if C is always larger than, $U_0 + u_1$ GSPR should never be built; and, that $U_0 + u_1$ is always larger than C is impossible.

We introduce the multiplier $k(0 < k \leq 1)$ to further clarify the difference between the concepts of GSPR as public good and that of the GSPR drawing right. If there exists and only exists one K so that the cost of the GSPR as public good is

kC(M) and that of the GSPR drawing right cost is (1-k)C(M), the GSPR as public good and the GSPR drawing right are clearly separated. Under this circumstance, B_i becomes the coordinate of i's share of the public good cost and that of the drawing right cost, β_i, b_i, subject to $\sum \beta_i = 1$ and $\sum b_i = 1$.

Many examples, such as panic buying of fuel in 1973 oil crisis [6], panic buying of fuel in Hurricane Katrina [7], panic buying of salt in Japan's nuclear crisis [8] and so on, suggest that consumer hoarding (or panic buying) may happen at supply interruption. According to the economic explanation given by [9], panic buying can be interpreted as the distortion of demand curve. When supply interruption happens, demanders are expected to value the supply unusually high and get less elastic to price, hence the demanders may suffer greater surplus loss. For the possibility of panic buying, assumption 7 is arranged.

Assumption 7: The announced supply interruption distorts the price elasticity of the aggregate demand. The degree of the distortion is decreasing on the size of the strategic inventory held at the immediate convenience of the consumers.

The Proposal

Suppose that the oil supply interruption is expected to take pace at time t and policies of GSPR are required to make at time 0. Currently, no GSPR drawing right has been pre allocated and the GSPR oil is sold to the market at, P_t the instant market price of time t, which means at time 0, k is regarded as 1. In other words, the current government is typically the kind of G_0. Our proposal is requiring the government to transform from G_0 to G_1.

More specifically, consumers and G_1 at time 0 are proposed to sign a contract which specifies: at time t consumer i is allowed to buy λ_i amount of GSPR stock at the price, $P_0 e^{rt}$ where P_0 is the market price at the contracting time point, provided that at time 0 consumer i pays a fair part of the total GSPR cost. To differentiate the GSPR under G_1 from the normal commercial oil stocks, we emphasize that the drawing rights also can only be executed at the government announced oil supply interruption.

THE PROOF OF THE EFFICIENCY IMPROVEMENT

Suppose at supply interruption, the oil supply is suddenly reduced to S' from the normal level S, and the size of the GSPR is M. We can do a geometric analysis and see there is a potential improvement between the ex ante and the ex post situations. In Figure 1, the abscissa stands for the amount of oil the consumers buy and the ordinate stand for the oil price; a marks point(q_2,g), b marks point (q_1,h), c marks point(q, t), d marks point (q_1,m), and e marks point (q_2,n).

In ex ante situation, the demanders have a total quantity M of GSPR drawing rights in hands and they can use it once the government declares a severe oil supply interruption. In this situation, the aggregate inverse demand function is distorted to $P_1(Q)$, and the consumers buy oil of amount q_1 need pay price ι averagely. But, at normal state the inverse demand function is $P_1(Q)$ (or D in Figure 1) and the consumers only need to pay the average price e for the same amount of oil. Thus, the consumers' surplus loss of is the size of trapezoid hbcl plus the size of curved triangle bcd, or. $(h-m)q_1 - \int_q^{q_1} DdQ$

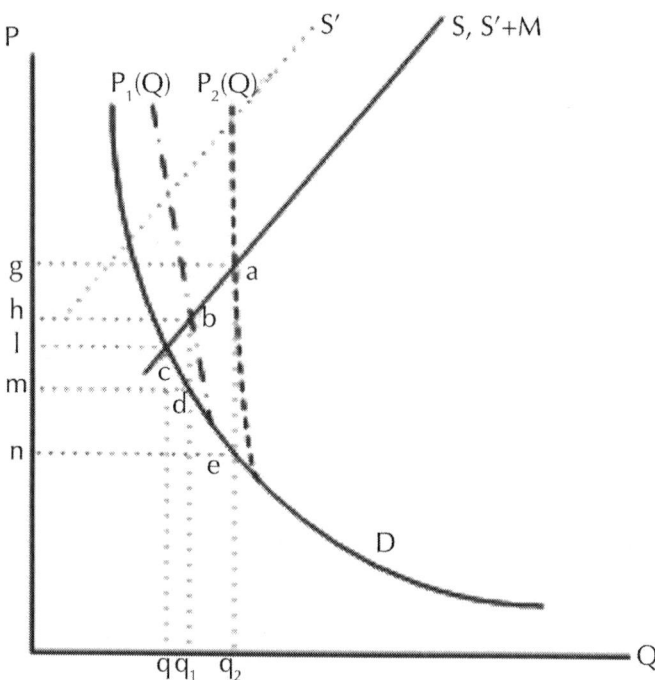

Figure 1: Ex ante preparation improves the GSPR by relieving panic buying.

In ex post situation, the demanders have little immediate strategic stockpile. In this situation, a round of chaotic panic buying would happen, and the aggregate inverse demand function would be distorted to $P_2(Q)$. The consumer surplus loss in ex post situation is measured by the size of trapezoid gacl plus the size of curved triangle ace, or.$(g-n)q_2 - \int_q^{q_1} DdQ$ Apparently, $(h-m)q_1 - \int_q^{q_1} DdQ$ is always less than $(g-n)q_2 -$

$\int_{q_1}^{q_2} DdQ$ hence the ex ante preparation is better than the ex post one in terms of efficiency.

THE PROOF OF THE INCENTIVE IMPROVEMENT

The General Proof

By incentive improvement we meant that under the proposed mechanism, consumers will enable the government to build a lager GSPR by contributing more. Under G_1, there exists one and only one k(0<k<1) so that $\partial U_0/\partial M = k \partial C/\partial M$ and $\partial U1/\partial M = (1-k)\partial C/\partial M$. Given the GSPR of size M, all consumers will pay less for GSPR as public good for KC(M) is definitely less than C(M). Therefore, the aggregate marginal utility curve under G_1 intersects the marginal cost curve to the right (on the vertical line $M=M_1$) of the point where the marginal utility curve under G_0 intersects the marginal cost (on the vertical line $M=M_1$), as Figure 2 shows. Thus, the government G_1 is required to build a GSPR of size M_1, and the total expenditure is $C(M_1)$ which is obviously larger than $C(M_0)$. Thus, the incentive improvement has easily been proved.

From Figure 2, it can also be easily inferred that the significance of the incentive improvement depends on the scale of k: the smaller the k the larger the M_1, which means if the consumers value GSPR drawing rights more, the incentive improvement would be more significant.

An Example

In order to deepen the understanding of the incentive improvement, we proceed with a simple numerical example.

Suppose C (M)=6M-1/(M+1)+1. *I*=3, and respectively the consumers' utility functions are:

$$u_1 = \ln(M+1) - 1/(M+1) + 1$$
$$u_{1d}(\lambda_1) = 2\ln(\lambda_1+1) - 1/(\lambda_1+1)$$
$$u_2 = 2\ln(M+1) - 2/(M+1) + 2$$
$$u_{2d}(\lambda_2) = 3\ln(\lambda_2+1) - 1/(\lambda_2+1) + 1$$
$$u_3 = 3\ln(M+1) - 3/(M+1)$$
$$u_{3d}(\lambda_3) = 4\ln(\lambda_3+1) - 1/(\lambda_3+1) + 1$$

It can be inferred that, under this market the aggregate utility function of the GSPR as public good is $U_0(M)=6\ln(M+1)-6/(M+1)+6$, and that of the GSPR drawing right is $U_1(M)=9\ln(M/3+1)-3/(M/3+1)+3$

Under G_0, k=1, and the optimal M_0 is determined by Equation (1) according to the principle "marginal utility equals marginal cost"

$$5/(M_0+1)^2 + 6(M_0+1) - 6 = 0 \tag{1}$$

and, the cost distribution among the consumers is determined by Equation (2).

$$\begin{cases} 5/(M_0+1)^2 + 6(M_0+1) - 6 = 0 \\ 1/(M_0+1) + 1/(M_0+1)^2 = \beta_1\left[6 + 1/(M_0+1)^2\right] \\ 2/(M_0+1) + 2/(M_0+1)^2 = \beta_2\left[6 + 1/(M_0+1)^2\right] \\ 3/(M_0+1) + 3/(M_0+1)^2 = \beta_3\left[6 + 1/(M_0+1)^2\right] \\ \beta_1 + \beta_2 + \beta_3 = 1 \end{cases} \tag{2}$$

By solving the combination of Equation (1) and Equation (2), we can obtain that, under G_0

,M_0=0.54 , the total expenditure on the GSPR is 3.60, and the cost distribution is $\beta(\beta_1, \beta_2, \beta_3)=(0.167, 0.33, 0.5)$ in terms of proportions, or the consumers need pay 0.60, 1.20 and 1.80 respectively. Under G_1.0<k<1, and the optimal M_1 is determined by Equation (3).

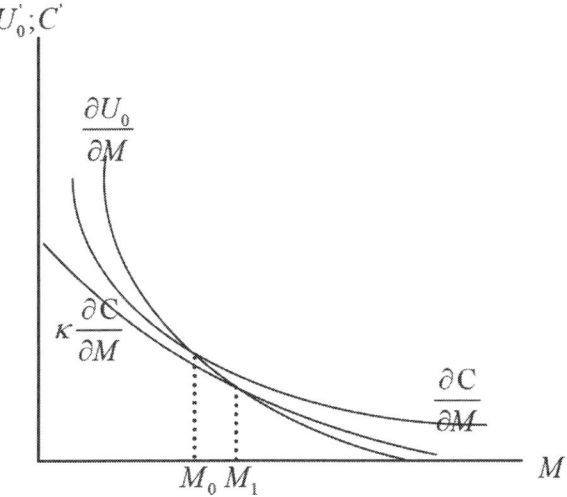

Figure 2. The intersections of the aggregate marginal utility curve and the marginal cost curves under G_0 and G_1.

$$\frac{6}{M_1+1}+\frac{6}{(M_1+1)^2}+\frac{9}{M_1+3}+\frac{9}{(M_1+3)^2}=6+\frac{1}{(M_1+1)^2} \tag{3}$$

and, the cost distribution among the consumers is determined by Equation (4).

$$\begin{cases} 1/(M_1+1)+1/(M_1+1)^2 = \beta_1 k\left[6+1/(M_1+1)^2\right] \\ 2/(M_1+1)+2/(M_1+1)^2 = \beta_2 k\left[6+1/(M_1+1)^2\right] \\ 3/(M_1+1)+3/(M_1+1)^2 = \beta_3 k\left[6+1/(M_1+1)^2\right] \\ 2/(\lambda_1+1)+1/(\lambda_1+1)^2 = b_1(1-k)\left[6+1/(M_1+1)^2\right] \\ 3/(\lambda_1+1)+1/(\lambda_1+1)^2 = b_2(1-k)\left[6+1/(M_1+1)^2\right] \\ 4/(\lambda_1+1)+1/(\lambda_1+1)^2 = b_3(1-k)\left[6+1/(M_1+1)^2\right] \\ \beta_1+\beta_2+\beta_3 = M_1 \\ b_1+b_2+b_3 = 1 \end{cases} \tag{4}$$

By solving the combination of Equation (3) and Equation (4), we can obtain that, under G_1, $M_1=1.351$, $k=0.55$, the total expenditure on the GSPR as public good is 4.81 and the total expenditure on the GSPR drawing rights is 3.86, and the cost distribution of the GSPR as public good is (0.167, 0.33, 0.5) in terms of proportions or (0.80, 1.60, 2.41) in terms of monetary values; the cost distribution of the GSPR drawing rights is (0.175, 0.337, 0.488) in terms of proportions or (0.68, 1.30, 1.89) in terms of monetary values; and,. $(\lambda_1, \lambda_2, \lambda_3)=(0.236. 0.455, 0.660)$ We can see that there has been a very significant improvement since M_1 is more the twice of M_0 and all together the consumers will spend more income under G_1 then the twice of that under G_0.

DISCUSSION

Sections 3 and 4 have theoretically proved that the proposal is supposed to improve the current GSPR, this improvement has practical difficulty though. To be optimal, balanced and impartial, the government G_1 shall make sure that, given M_1, $\Sigma b_i = 1$ on one side, and $\Sigma \lambda_i = M_1$ on the other. Those equalities require that each consumer reports to G_1 his/her utilities functions honestly. We have assumed that under both G_0 and G_1, the consumers give up their propensity to free-ride on others. This assumption doesn't reflect the possibility that G_1 would reinforce the consumers' propensity to free-ride, since some consumers may think the others have stronger desire for the GSPR drawing rights, and even they choose to pay nothing a GSPR will be financed. However, it's really more difficult to solve the free-riding problem under G_0 than G_1? This question needs in-depth investigation into the consumers' behavioral motives and the legitimacy of the governmental authority to answer, which transcends the scope of this paper. Nevertheless, the proposal reserves a great possibility to improve the current GSRP as the example suggests, hence still worth serious considering.

CONCLUSIONS

GSPR programs are supposed to save the probable larger loss with a confirmed cost, so the efficiency of them is very important. For historical events have suggested that panic buying is quite likely to happen at supply interruptions, we argue that any in-advance preparation for emergencies should have better immediacy. GSPR can be more immediate if its drawing rights have been pre-allocated. Moreover, this pre-allocation may greatly improve the society's incentive to finance GSPR. Therefore, we propose that when deciding the size and the cost distribution of GSPR at time 0, the GSPR drawing rights shall be allocated simultaneously. The GSPR drawing rights specify that if an oil supply interruption happens at time, t consumer i who has bought the GSPR drawing right of λ_i is allowed to buy the released GSPR oil of amount λ_i at the preset price $P_0 e^{rt}$, where P_0 is the market price at time 0. Of course, the GSPR drawing rights can only be executed at the government that declares oil supply interruption.

ACKNOWLEDGEMENTS

This paper was founded by the Ministry of Education of the People's Republic of China under the name of its Philosophy and Social Science Major Issue Research Project "Research on Expanding the Oil & Gas Strategic Reserve" (Grant No. 11JZD048) and National Natural Science Foundation of China under the name of its project "Research on the System of Natural Gas Security & Emergency Measures" (Grant No. 71273277). We owe the greatest thanks to those two founders. However, we have to say at all this paper doesn't necessarily represent the opinion of those two founders.

REFERENCES

1. Z. M. Liu, L. W. Zhu and J. M. Wang, "Optimization of China's Strategic Petroleum Reserve Policy: A Markovian Decision Approach," Computers & Industrial Engineering, Vol. 63, No. 3, 2012, pp. 626-633. http://dx.doi.org/10.1016/j.cie.2011.10.014
2. International Energy Agency, "IEA Response System for Oil Supply Emergencies," 2012 Edition, 2012. http://www.iea.org/publications/freepublications/publication/name,3714,en.html
3. J. Taylor and P. V. Doren, "The Case against the Strategic Petroleum Reserve," Cato Institute, Washington DC, 2005.
4. P. A. Samuelson, "The Pure Theory of Public Expenditure," The Review of Economics and Statistics, Vol. 36, No. 4, 1954, pp. 387-389. http://dx.doi.org/10.2307/1925895

5. D. Holtz-Eakin and M. E. Lovely, "Scale Economies, Returns to Variety, and the Productivity of Public Infrastructure," Regional Science and Urban Economics, Vol. 26, No. 2, 1996, pp. 105-123. http://dx.doi.org/10.1016/0166-0462(95)02126-4
6. S. Cooper, "A Tale of Two Oil Shocks. Part 1: 1973- 76," Frontline, 2012. http://www.pbs.org/wgbh/pages/frontline/tehranbureau/2012/06/a-tale-of-two-oil-shocks-part-1-1973-
7. E. Chow and J. Elkind, "Hurricane Katrina and US Energy Security," Survival, Vol. 47, No. 4, 2005, pp. 145- 160. http://dx.doi.org/10.1080/00396330500433449
8. J. C. Wei, L. Zhou and X. Zhou, "Analysis of the Evolution Mechanism of Mass Panic Buying under the Public Crisis. Case Study on Salt Panic Buying in China during the Japan Nuclear Crisis," Journal of Management Case Studies, Vol. 4, No. 6, 2011, pp. 478-486.
9. S. Ronald, J. Keith and A. T. Khairy, "Scarcity and Hoarding: Economic and Social Explanations and Marketing Implications," Advances in Consumer Research, Vol. 2, 1975, pp. 203-216.

Chapter 7

Studies on the Evaporation Regulation Mechanisms of Crude Oil and Petroleum Products

Merv F. Fingas

Spill Science, Edmonton, Canada

ABSTRACT

Various concepts for oil evaporation prediction are summarized. Models can be divided into those models that use the basis of air-boundary-regulation or those that do not. Experiments were conducted to determine if oil and petroleum evaporation is regulated by the saturation of the air boundary layer. Experiments included the examination of the evaporation rate with and without wind, in which case it was found that evaporation rates were similar for all wind conditions and no-wind conditions. Experiments where the area and mass varied showed that boundary-layer regulation was not governing for petroleum products. Under all experimental and environmental conditions, oils or petroleum products were not found to be boundary-layer regulated. Experiments on the rate of evaporation of pure compounds showed that compounds larger than Decane were not boundary-layer regulated. Many oils and petroleum products contain few compounds smaller than decane, and this explains why their evaporation is not air boundary-layer limited. Comparison of the air saturation levels of various oils and petroleum products shows that the saturation concentration of water, which is strongly air boundary-regulated, is significantly less than that of several petroleum hydrocarbons. Lack of air boundary-layer regulation for oils is shown to be a result of both this higher saturation concentration as well as a low (below boundary-layer value) evaporation rate.

INTRODUCTION

Evaporation is an important process for most oil spills. In a few days, typical crude oils can lose up to 45% of their volume. The Macondo oil lost up to 60% in a short time when released under water at high pressure [1]. Almost all oil spill models include evaporation as a process and output of the model. Evaporation plays a prime role in the fate of most oils. Many crude oils must undergo evaporation before they will form water-in-oil emulsions [1]. Light oils will change very dramatically from fluid to viscous. Heavy oils will become solid-like. Many oils after long evaporative exposure form tar balls or heavy tar mats. Despite the importance of the process, little work has been conducted on the basic physics and chemistry of oil spill evaporation [2]. The difficulty with studying oil evaporation is that oil is a mixture of hundreds of compounds and oil composition varies from source to source and even over time. Much of the work described in the older literature focuses on calibrating equations developed for water evaporation [2].

The mechanisms that regulate evaporation are important [3,4]. Evaporation of a liquid can be considered as the movement of molecules from the surface into the vapour phase above it. The immediate layer of air above the evaporation surface is known as the air boundary layer [5].

This boundary layer is the intermediate interface between the air and the liquid and might be viewed as very thin such as less than one mm. The characteristics of this air boundary layer can influence evaporation. In the case of water, the boundary layer regulates the evaporation rate. Air can hold a variable amount of water, depending on temperature, as expressed by the relative humidity. Under conditions where the air boundary layer is not moving (no wind) or has low turbulence, the air immediately above the water quickly becomes saturated and evaporation slows. The actual evaporation of water proceeds at a small fraction of the possible evaporation rate because of the saturation of the boundary layer. The air-boundary-layer physics is then said to regulate the evaporation of water. This regulation manifests as the increase of evaporation with wind or turbulence. When turbulence is weak, evaporation can slow down by orders-of-magnitude. The molecular diffusion of water molecules through air is at least 10^3 times slower than turbulent diffusion [5]. If the evaporation of oil was like that of water and was air boundary-layer regulated, one could write the mass transfer rate in semi-empirical form (also in generic and unitless form) as:

$$E = KCT_u S \tag{1}$$

where E is the evaporation rate in mass per unit area, K is the mass transfer rate of the evaporating liquid, presumed constant for a given set of physical conditions, sometimes denoted as k_g (gas phase mass transfer coefficient, which may incorporate some of the other parameters noted here), C is the concentration (mass) of the evaporating fluid as a mass per volume, T_u is a factor characterizing the relative intensity of turbulence, S is a factor that relates to the saturation of the boundary layer above the evaporating liquid. The saturation parameter, S, represents the effects of local advection on saturation dynamics. If the air is already saturated with the compound in question, the evaporation rate approaches zero. This also relates to the scale length of an evaporating pool. If one views a large pool over which a wind is blowing, there is a high probability that the air is saturated downwind and the evaporation rate per unit area is lower than for a smaller pool. It should be noted that there are many equivalent ways of expressing this fundamental evaporation equation. Much of the pioneering work for water evaporation work was performed by Sutton [6]. Sutton proposed the following equation based largely on empirical work:

$$E = KC_s U^{7/9} d^{-1/9} Sc^{-r} \tag{2}$$

where C_s is the concentration of the evaporating fluid (mass/volume), U is the wind speed, d is the area of the pool, Sc is the Schmidt number and r is the empirical exponent assigned values from 0 to 2/3. Other parameters are defined as above. The terms in this equation are analogous to the very generic equation (1), proposed above. The turbulence is expressed by a combination of the wind speed, U, and the Schmidt number, Sc. The Schmidt number is the ratio of kinematic viscosity of air (ν) to the molecular diffusivity (D) of the diffusing gas in air, i.e. a dimensionless expression of the molecular diffusivity of the evaporating substance in air. The coefficient of the wind power typifies the turbulence level. The value of 0.78 (7/9) as chosen by Sutton, represents a turbulent wind whereas a coefficient of 0.5 would represent a wind flow that was more laminar. The scale length is represented by d and has been given an empirical exponent of −1/9. This represents, for water, a weak dependence on size. The exponent of the Schmidt number, r, represents the effect of the diffusivity of the particular chemical, and historically was assigned values between 0 and 2/3 [5].

This expression for water evaporation was subsequently used by those working on oil spills to predict and describe oil and petroleum evaporation. Much of the literature follows the work of Mackay [7,8]. Mackay and Matsugu [7] corrected the equations for hydrocarbons using the evaporation rate of cumene. Data on the evaporation of water and cumene have been used to correlate the gas phase mass transfer coefficient as a function of wind-speed and pool size by the equation:

$$K_m = 0.0292 U^{0.78} X^{-0.11} Sc^{-0.67} \tag{3}$$

where K_m is the mass transfer coefficient in units of mass per unit time and X is the pool diameter or the scale size of evaporating area. Stiver and Mackay [8] subsequently developed this further by adding a second equation:

$$N = k_m A P / (RT) \tag{4}$$

where N is the evaporative molar flux (mol/s), k_m is the mass transfer coefficient at the prevailing wind (m/s), A is the area (m²), P is the vapour pressure of the bulk liquid (Pascals), R is the gas constant [8.314 Joules/ (mol-K)], and T is the temperature (K).

Thus, air boundary layer regulation was assumed to be the primary regulation mechanism for oil and petroleum evaporation. This assumption was never tested by experimentation, as revealed by a literature search [2]. The implications of these assumptions are that evaporation rate for a given oil is increased by:

- increasing turbulence
- increasing wind speed, and
- increasing the surface area of a given mass of oil.

These factors can then be verified experimentally to test if oil is boundary-layer regulated or not. These factors formed the basis of experimentation for this paper.

EXPERIMENTAL

Evaporation rate was measured by weight loss using an electronic balance. The balance was a Mettler PM4000. The weight was recorded using a laptop computer, a serial cable to the balance and the software program, "Collect" (Labtronics, Richmond, Ontario).

Measurements were conducted in the following fashion. A tared petri dish of defined size was loaded with a measured amount of oil. At the end of the experiment vessels were cleaned and rinsed with dichloromethane and a new experiment started. The weight loss dishes were standard glass petri dishes from Corning. A standard 139 mm diameter (ID) dish was used for most experiments. For the experiments in which area was a variable, dishes of other diameters were employed. Diameters and other dimensions were measured using a Mitutoyo digital vernier caliper. The lip, height of the dish above the oil, with the 139 mm dish varied from 2 to 10 mm depending on depth of fill. For the other dishes the lip varied from 2 to 20 mm.

Measurements were done in one of three locations; inside a fume hood, inside a controlled temperature room, or on a counter top. Some experiments were conducted in the fume hood, where there was no temperature regulation. Temperatures were measured using a Keithley 871 digital thermometer with a thermocouple supplied by the same firm. Temperatures were taken at the beginning and the end of a given experimental run.

The constant temperature chamber (room) employed was a Constant Temperature model. It could maintain temperatures from −40°C to +60°C and regulate the chosen temperature within ±1°C.

In experiments involving wind, air velocities were measured using a Taylor vane anemometer and a Tadi, "Digital Pocket Anemometer". Measurements were taken at the closest position above the glass vessel floor and at the lip level. These velocities were later confirmed using a hot wire anemometer and appropriate data manipulations of the outputs. The anemometer was a TSI— Thermo Systems model 1053b, with power supply (TSI model 1051-1), averaging circuit (TSI model 1047) and signal linearlizing circuit (TSI model 1052). The voltage from the averaging circuit was read with a Fluke 1053 voltmeter. The hot wire sensor (TSI model 1213-60) was angled at 45°. The sensor probe resistance at 0°C was 7.21 ohms and the sensor was operated at 12 ohms for a recommended operating temperature of 250°C. Data from the hot wire anemometer was collected on a Campbell Scientific CR-10 data logger at a rate of 64 Hz.

Evaporation data were collected on a laptop computer and subsequently transferred to other computers for analysis. The "Collect" program records time and the weight directly. Data were recorded in ASCII format and converted to Excel format. Curve fitting was performed using the software program "Table Curve", Jandel Scientific Corporation, San Raphael, California.

Oils were taken from supplies of Environment Canada and were supplied by various oil companies for environmental testing. Table 1 lists the properties and descriptions of the test liquids [9].

RESULTS AND DISCUSSION

Table 2 lists the experiments performed and the results in terms of the best fit equations. These were done by curve fitting using the program Table Curve, as noted above. The best fit was done on the basis of the simplest equation fitting with the highest regression coefficient (R^2). The results are presented in the order of the experimental series:

Wind Experiments

Experiments on the evaporation of oil with and without wind, were conducted with three oils, ASMB (Alberta Sweet Mixed Blend crude oil), Gasoline, FCC Heavy Cycle (a processed oil), and with water. Water formed a baseline data set since much is known about its evaporation behaviour [3,4]. Regressions on the data were performed and the equation parameters calculated, are shown in Table 3. Curve coefficients are the constants from the best fit equation (Evap = a ln(t), t = time in minutes, for logarithmic equations or Evap = $a\sqrt{t}$, for the square root equations).

Table 1: Properties of the test liquids

Test Liquid	Description	Densit) g/inl.	Boiling Point °C
ASMB	Alberta Sweet Mixed Blend—A common crude oil in Canada	0.839	initial-37
Water		1	100
FCC-heavy	A highly-cycled refinery intermediate containing few components	0.908	
Gasoline	Standard automotive gasoline	0.709	initial-5
Benzene	Pure Hydrocarbon C6	0.879	80.1
Dodecanc	Pure Hydrocarbon ClO	0.749	213
Undecane	Pure Hydrocarbon C 1 l	0.742	196
p-Xylenc	Pure Hydrocarbon C8	0.861	139
Nonanc	Pure Hydrocarbon C9	0.722	151
Decanc	Pure Hydrocarbon ClO	0.73	174
Heptonc	Pure Hydrocarbon C7	0.684	98
Octane	Pure Hydrocarbon C8	0.703	126
Decahydron	Decahydronaphthalene pure hydrocarbon ClO	0.893	195
Tridecanc	Pure Hydrocarbon C13	0.755	226
Hexadecane	Pure Hydrocarbon C 16	0.773	287

Table 2: Experimental summary

Number	Experimental Purpose	Oil Type	Total Time (hr)	Pan (cm²) Area	Initial (mm) Thickness	Temp °C	Wind m/s	Variable	Variable Value	R^2 Best Equation	Best Equation
1	Thickness	ASMB	15	151	0.65	21.2	0	thick	0.65	0.991	ln
2	Thickness	ASMB	22	268	0.72	21	0	thick	0.72	0.978	ln
3	Thickness	ASMB	23	270	1.3	21.8	0	thick	1.3	0.97	ln
4	Thickness	ASMB	182	151	0.63	22.6	0	thick	0.63	0.99	ln
5	Thickness	ASMB	15	151	1.59	22.4	0	thick	1.59	0.937	ln
6	Thickness	ASMB	51	151	1.78	21.9	0	thick	1.78	0.975	ln
7	Thickness	ASMB	65	151	2.14	24.4	0	thick	2.14	0.954	ln
8	Thickness	ASMB	25	151	2.69	23.8	0	thick	2.69	0.952	ln
9	Thickness	ASMB	73	151	2.84	21.7	0	thick	2.84	0.96	ln
10	Thickness	ASMB	36	151	4.55	22.8	0	thick	4.55	0.963	ln
11	Thickness	ASMB	18	151	9.08	20.1	0	thick	9.08	0.879	ln
12	Thickness	ASMB	73	151	7.61	20.3	0	thick	7.61	0.886	ln
13	Thickness	ASMB	217	151	5.21	20	0	thick	5.21	0.937	ln
14	Thickness	ASMB	64	151	1.53	22.1	0	thick	1.53	0.981	ln
15	Thickness	ASMB	56	151	3.21	17.8	0	thick	3.21	0.952	ln
16	Thickness	ASMB	47	151	1.33	19.2	0	thick	1.33	0.987	ln
17	Thickness	ASMB	23	151	0.59	18.8	0	thick	0.59	0.988	ln
18	Thickness	ASMB	25	151	0.63	20.1	0	thick	0.63	0.985	ln
19	Thickness	ASMB	71	151	1.96	23.1	0	thick	1.96	0.976	ln
20	Thickness	ASMB	32	151	2.54	18.6	0	thick	2.54	0.977	ln
21	Thickness	ASMB	89	151	5.27	22.9	0	thick	5.27	0.98	ln
22	Thickness	ASMB	76	151	1.43	20.4	0	thick	1.43	0.993	ln
23	Thickness	ASMB	66	151	1.39	20.3	0	thick	1.39	0.986	ln
24	Thickness	ASMB	88	151	2.8	19.1	0	thick	2.8	0.962	ln
25	Area	ASMB	50	16	7.45	24.2	0	area	16 cm²	0.969	ln
26	Area	ASMB	25	16	3.72	23.9	0	area	16 cm²	0.96	ln
27	Area	ASMB	21	16	1.58	8	0	area	16 cm²	0.72	ln
28	Area	ASMB	25	16	0.79	24.6	0	area	16 cm²	0.791	ln
29	Area	ASMB	50	62	3.84	22.5	0	area	62 cm²	0.992	ln
30	Area	ASMB	22	62	1.92	15.6	0	area	62 cm²	0.996	ln
31	Area	ASMB	26	62	1.58	25.3	0	area	62 cm²	0.982	ln
32	Area	ASMB	23	62	0.79	23.8	0	area	62 cm²	0.994	ln
33	Area	ASMB	24	161	1.48	21	0	area	161 cm²	0.987	ln
34	Area	ASMB	23	161	0.79	25.2	0	area	161 cm²	0.973	ln
35	Area	ASMB	50	161	1.58	23.9	0	area	161 cm²	0.941	ln

36	Area	ASMB	83	161	3.7	19.1	0	area	161 cm^2	0.933	ln
37	Area	ASMB	50	161	2.22	21	0	area	161 cm^2	0.99	ln
38	Area	ASMB	25	161	0.74	20	0	area	161 cm^2	0.953	ln
39	Area	ASMB	74	206	1.58	18	0	area	206 cm^2	0.984	ln
40	Area	ASMB	20	206	0.79	21	0	area	206 cm^2	0.974	ln
41	Area	ASMB	51	206	1.16	19.5	0	area	206 cm^2	0.963	ln
42	Area	ASMB	44	151	1.58	20.5	0	area	151 cm^2	0.993	ln
43	Area	ASMB	26	151	0.79	19	0	area	151 cm^2	0.994	ln
44	Wind	ASMB	23	151	1.58	22.9	1.45	wind	1.0 m/s	0.98	ln
45	Wind	ASMB	24	151	1.58	22	1.45	wind	1.0 m/s	0.972	ln
46	Wind	ASMB	42	151	3.16	21.1	1.45	wind	1.0 m/s	0.99	ln
47	Wind	ASMB	46	151	3.16	21.2	1.45	wind	1.0 m/s	0.993	ln
48	Wind	Water	3	151	1.32	21.8	1.45	wind	1.0 m/s	0.997	lin
49	Wind	Water	3	151	1.32	21.8	1.45	wind	1.0 m/s	0.997	lin
50	Wind	Water	3	151	2.65	21.8	1.45	wind	1.0 m/s	0.999	lin
51	Wind	ASMB	21	151	1.58	22.1	1.65	wind	1.6 m/s	0.981	ln
52	Wind	ASMB	22	151	1.58	21.4	1.65	wind	1.6 m/s	0.949	ln
53	Wind	ASMB	23	151	1.58	21.4	1.65	wind	1.6 m/s	0.996	ln
54	Wind	ASMB	46	151	3.16	22.7	1.65	wind	1.6 m/s	0.986	ln
55	Wind	ASMB	20	151	1.58	22.8	1.65	wind	1.6 m/s	0.977	ln
56	Wind	Water	1	151	1.32	21.7	1.65	wind	1.6 m/s	0.998	lin
57	Wind	ASMB	17	151	1.58	23.9	1.65	wind	1.6 m/s	0.978	ln
58	Wind	Water	3	151	1.32	22.2	1.65	wind	1.6 m/s	0.999	lin
59	Wind	Water	5	151	2.65	23.6	1.65	wind	1.6 m/s	0.989	lin
60	Wind	ASMB	22	151	1.58	24.3	1.65	wind	1.6 m/s	0.981	ln
61	Wind	Water	1	151	1.32	23.4	1.85	wind	2.1 m/s	0.998	lin
62	Wind	ASMB	44	151	3.16	23	1.85	wind	2.1 m/s	0.991	ln
63	Wind	ASMB	6	151	1.58	21.7	1.85	wind	2.1 m/s	0.993	ln
64	Wind	ASMB	39	151	3.16	20.4	1.85	wind	2.1 m/s	0.993	ln
65	Wind	Water	2	151	1.32	21.8	1.85	wind	2.1 m/s	0.994	lin
66	Wind	Water	5	151	2.65	22.6	1.85	wind	2.1 m/s	0.998	lin
67	Wind	ASMB	12	151	1.58	22.4	1.85	wind	2.1 m/s	0.993	ln
68	Wind	FCC-heavy	32	151	2.92	21.7	1.85	wind	2.1 m/s	0.987	sq. rt.
69	Wind	Gasoline	1	151	1.87	22.6	1.85	wind	2.1 m/s	0.983	ln
70	Wind	Gasoline	2	151	3.74	22.4	1.85	wind	2.1 m/s	0.975	ln
71	Wind	FCC-heavy	22	151	1.46	22.3	1.85	wind	2.1 m/s	0.996	sq. rt.
72	Wind	ASMB	21	151	1.58	23.4	3.8	wind	2.5 m/s	0.981	ln

73	Wind	Water	1	151	1.32	22.4	3.8	wind	2.5 m/s	0.997	lin
74	Wind	Water	2	151	2.65	22.2	3.8	wind	2.5 m/s	0.999	lin
75	Wind	Gasoline	0	151	1.87	22.2	3.8	wind	2.5 m/s	0.984	ln
76	Wind	Gasoline	1	151	3.74	21.9	3.8	wind	2.5 m/s	0.994	ln
77	Wind	Water	3	151	1.32	21.7	0	wind	0	0.999	lin
78	Wind	FCC-heavy	47	151	2.92	21.4	3.8	wind	2.5 m/s	0.994	sq. rt.
79	Wind	FCC-heavy	39	151	1.46	22	3.8	wind	2.5 m/s	0.997	sq. rt.
80	Wind	ASMB	34	151	1.58	22.5	3.8	wind	2.5 m/s	0.993	ln
81	Wind	ASMB	18	151	3.16	21	3.8	wind	2.5 m/s	0.997	ln
82	Wind	Water	1	151	1.32	22	3.8	wind	2.5 m/s	0.986	lin
83	Wind	Water	2	151	2.65	22.9	3.8	wind	2.5 m/s	0.994	lin
84	Wind	FCC-heavy	19	151	1.46	23	3.8	wind	2.5 m/s	0.992	sq. rt.
85	Wind	Gasoline	1	151	1.87	22.1	1.65	wind	1.6 m/s	0.996	ln
86	Wind	Gasoline	3	151	3.74	22.4	1.65	wind	1.6 m/s	0.983	ln
87	Wind	FCC-heavy	40	151	2.92	22.3	1.65	wind	1.6 m/s	0.997	sq. rt.
88	Wind	Gasoline	1	151	1.87	21.8	1.45	wind	1.0 m/s	0.992	ln
89	Wind	Gasoline	2	151	3.74	22.1	1.45	wind	1.0 m/s	0.973	ln
90	Wind	FCC heavy	21	151	1.46	23.1	1.45	wind	1.0 m/s	0.99	sq. rt.
91	Wind	FCC heavy	51	151	2.92	24.2	1.45	wind	1.0 m/s	0.996	sq. rt.
92	Wind	FCC heavy	46	151	1.46	24	0	wind	0	0.986	sq. rt.
93	Wind	Water	3	151	1.32	23.9	0	wind	0	0.999	lin
94	Wind	FCC heavy	87	151	2.92	23.9	0	wind	0	0.996	ln
95	Wind	Water	8	151	2.65	25	0	wind	0	0.999	lin
96	Wind	Water	16	151	2.65	25.1	0	wind	0	0.998	lin
97	Wind	Gasoline	7	151	1.87	22.5	0	wind	0	0.92	ln
98	Wind	Gasoline	17	151	3.74	22.5	0	wind	0	0.944	ln
99	Wind	Water	6	151	1.32	23	0	wind	0	0.99	lin
100	Pure cmpd.	Benzene	2	151	1.51	23.9	0	rate		0.999	lin
101	Pure cmpd.	Dodecane	45	151	1.77	23.3	0	rate		0.999	lin
102	Pure cmpd.	Undecane	46	151	1.79	24.3	0	rate		0.999	lin
103	Pure cmpd.	p-Xylene	7	151	1.54	24	0	rate		0.989	lin
104	Pure cmpd.	Nonane	11	151	1.83	24	0	rate		0.999	lin
105	Pure cmpd.	Decane	19	151	1.81	22.3	0	rate		0.998	lin
106	Pure cmpd.	Heptane	3	151	1.94	18.5	0	rate		0.999	lin
107	Pure cmpd.	Octane	3	151	1.88	20.4	0	rate		0.997	lin
108	Pure cmpd.	Decahydronapthalene	18	151	1.48	21	0	rate		0.996	lin
109	Pure cmpd.	Tridecane	23	151	1.79	21.1	0	rate		0.986	lin
110	Pure cmpd.	Hexadecane	167	151	1.71	15	0	rate		0.847	lin

Table 3: Data from the wind tests

Type	Loading grams	Curve Coefficients* % evap	Curve Coefficients* Abs. Wt.	Wind m/s	Type	Loading grams	Curve Coefficients % evap	Curve Coefficients Abs. Wt.	Wind m/s
ASMB	20	4.22	0.844	0	FCC heavy**	20	0.414	0.117	0
ASMB	20	5.28	1.06	1	FCC heavy	20	0.887	0.178	1
ASMB	20	5.3	1.06	1	FCC-heavy	20	0.8	0.161	2.1
ASMB	20	5.19	1.04	1.6	FCC-heavy	20	1.13	0.225	2.5
ASMB	20	5.27	1.05	1.6	FCC-heavy	20	0.905	0.181	2.5
ASMB	20	5.15	1.03	1.6					
ASMB	20	5.63	1.13	1.6	FCC heavy	20	0.414	0.2	0
ASMB	20	5.47	1.09	1.6	FCC heavy	40	0.66	0.264	1
ASMB	20	5.54	1.11	1.6	FCC-heavy	40	0.669	0.268	1.6
ASMB	20	5.78	1.16	2.1	FCC-heavy	40	0.557	0.223	2.1
ASMB	20	5.52	1.11	2.1	FCC-heavy	40	0.785	0.314	2.5
ASMB	20	5.82	1.16	2.5					
ASMB	20	5.52	1.1	2.5	Gasoline	20	12.2	3.36	0
					Gasoline	20	19.5	3.9	1
ASMB	40	4.09	2	0	Gasoline	20	19.7	3.93	1.6
ASMB	40	4.77	1.91	1	Gasoline	20	18.2	3.64	2.1
ASMB	40	4.77	1.91	1	Gasoline	20	21.6	4.32	2.5
ASMB	40	4.9	1.96	1.6					
ASMB	40	4.85	1.94	2.1	Gasoline	40	12.2	6	0
ASMB	40	4.99	2	2.1	Gasoline	40	16	6.4	1
ASMB	40	5.21	2.08	2.5	Gasoline	40	16.6	6.65	1.6
					Gasoline	40	15.4	6.15	2.1
Water	20	0.186	0.0372	0	Gasoline	40	16.6	6.64	2.5
Water	20	0.179	0.0357	0					
Water	20	0.178	0.0356	0	Water	40	0.088	0.0354	0
Water	20	0.592	0.118	1	Water	40	0.0778	0.0311	0
Water	20	0.612	0.112	1	Water	40	0.34	0.136	1
Water	20	0.512	0.102	1.6	Water	40	0.312	0.137	1.6
Water	20	0.515	0.103	1.6	Water	40	0.316	0.127	2.1
Water	20	0.7	0.14	2.1	Water	40	0.56	0.224	2.5
Water	20	0.603	0.12	2.1	Water	40	0.602	0.241	2.5
Water	20	1.02	0.206	2.5					
Water	20	1.04	0.209	2.5					

**The equations used for FCC Heavy are square root and for water, linear.

**The equations used for FCC Heavy are square root and for water, linear.

Data were calculated separately for percentage of weight lost and absolute weight. Both values show the small relative upward tendency with respect to wind effects. The plots of wind speed versus the evaporation rate (as a percentage of weight lost) for each oil type are shown in Figures 1 to 4. These figures show that the evaporation rates for oils and even the light products, gasoline and FCC Heavy Cycle, are not increased by a significant amount with increasing wind speed.

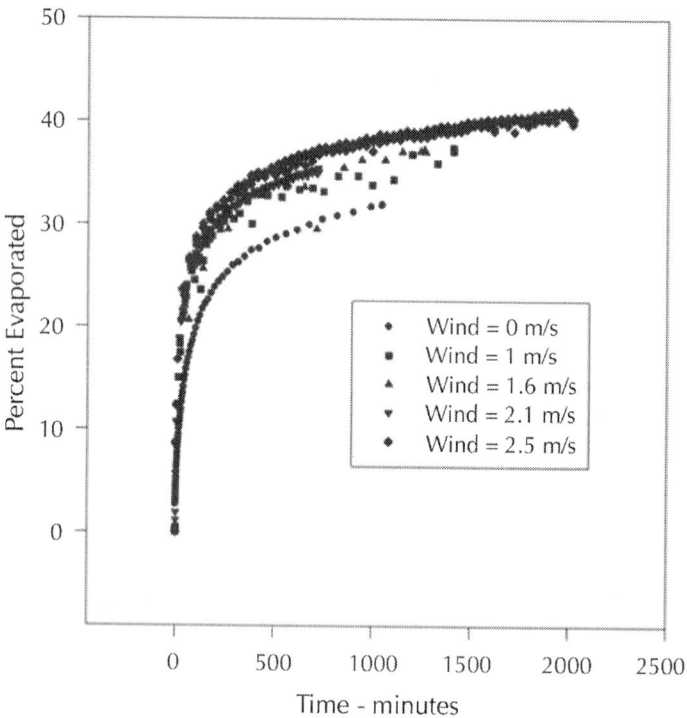

Figure 1: Evaporation of ASMB with varying wind velocities.

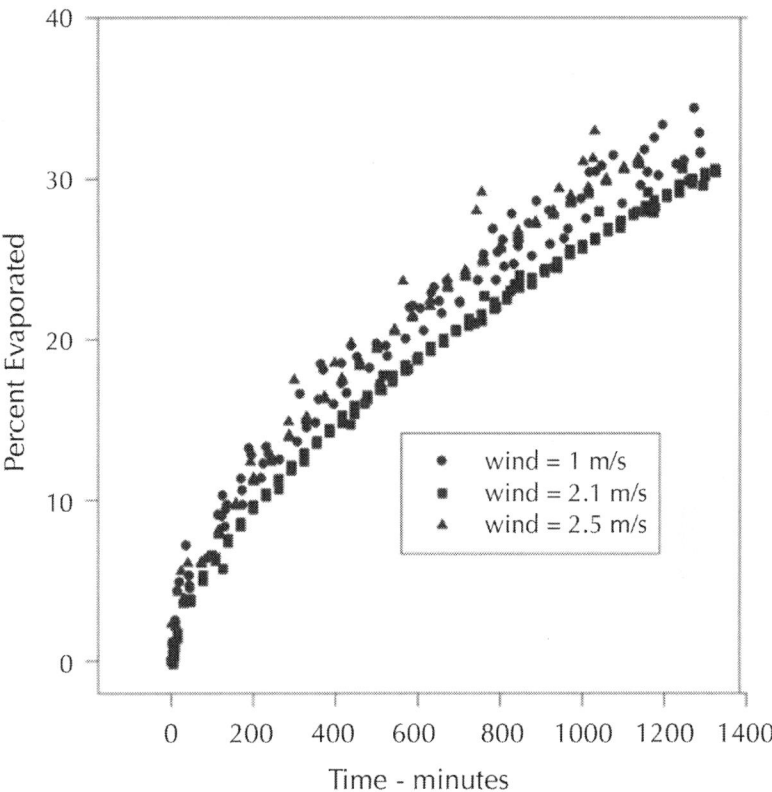

Figure 2: Evaporation of FCC-Heavy with varying wind velocities.

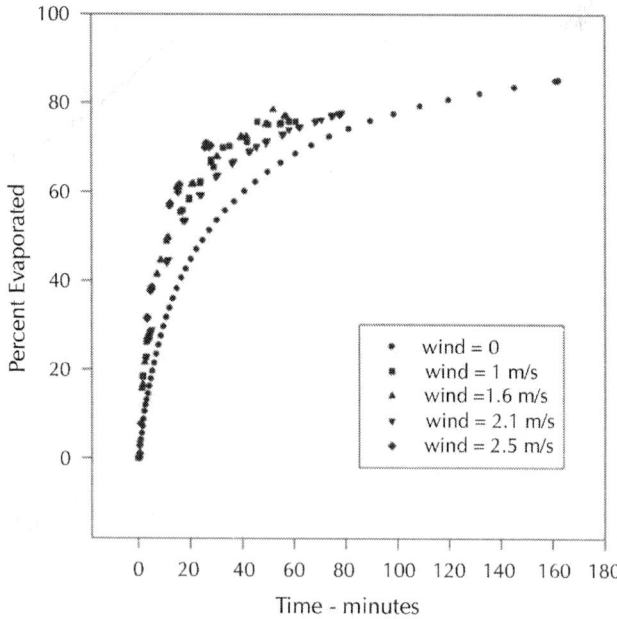

Figure 3: Evaporation of gasoline with varying wind velocities.

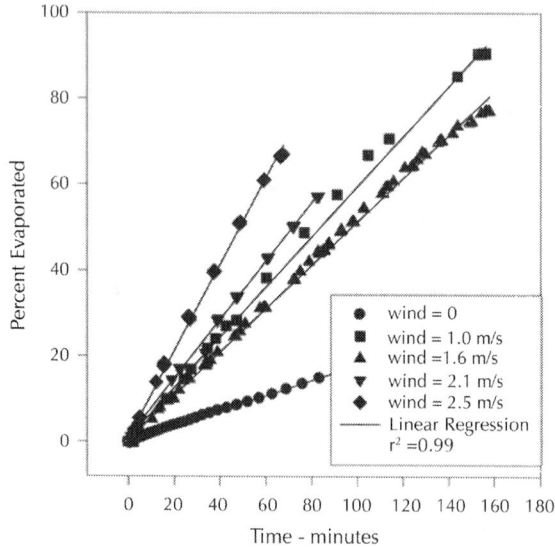

Figure 4: Evaporation of water (20 g) with varying wind velocities.

In some cases, there is a small rise from the 0-wind level to the 1-m/s level, but after that, the rate remains relatively constant. The evaporation rate after the 0-wind value is nearly identical for all oils. The oil evaporation data can be compared to the evaporation of water, as illustrated in Figure 4.

These data show the classical relationship of the water evaporation rate correlated with the wind speed (evaporation varies as $U^{0.78}$, where U is wind speed). This indicates that the oils used here are not boundary-layer regulated. Figure 5 shows the rates of evaporation compared to the wind speed for all the liquids used in this study. This figure shows the evaporation rates of all test liquids versus wind speed. The lines shown are those calculated by linear regression using the graphics software, SigmaPlot (Washington, DC). This clearly shows that water evaporation rate increased, as expected, with increasing wind velocity. The oils. ASMB, FCC heavy cycle and gasoline, do not show a measurable increase with increasing wind speed. In any case, the oils do not show the $U^{0.78}$ relationship that water shows.

All the above data show that oil is not boundary-layer regulated. Water shows the classic boundary-layer regulation.

Study of Mass and Evaporation Rate

ASMB oil was again used to conduct a series of experiments with volume as the major variant. Alternatively thickness and area were held constant to ensure that the strict relationship between these two variables did not affect the final regression results. Figure 6 illustrates the relationship between evaporation rate and volume of evaporation material (also equivalent to mass of evaporating material). This figure illustrates a strong correlation between oil mass (or volume) and evaporation rate. This suggests no air boundary-layer regulation is at work, since for an air boundary-layer regulated material evaporation is not affected by mass in the same area.

Study of the Evaporation of Pure Hydrocarbons—with and without Wind

A study of the evaporation rate of pure hydrocarbons was conducted to test the classic boundary-layer evaporation theory as applied to the hydrocarbon constituents of oils.

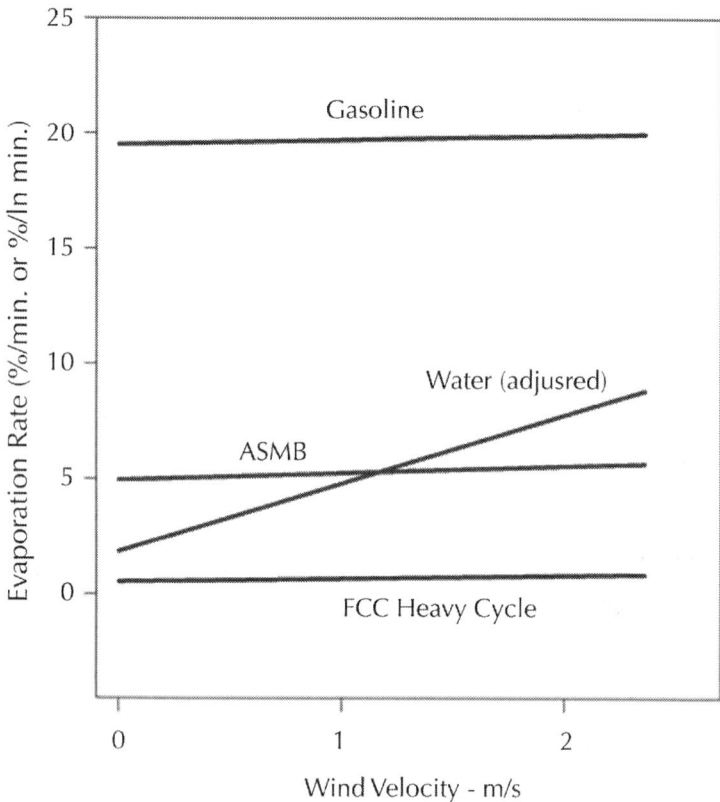

Figure 5: Correlation evaporation rates and wind velocities.

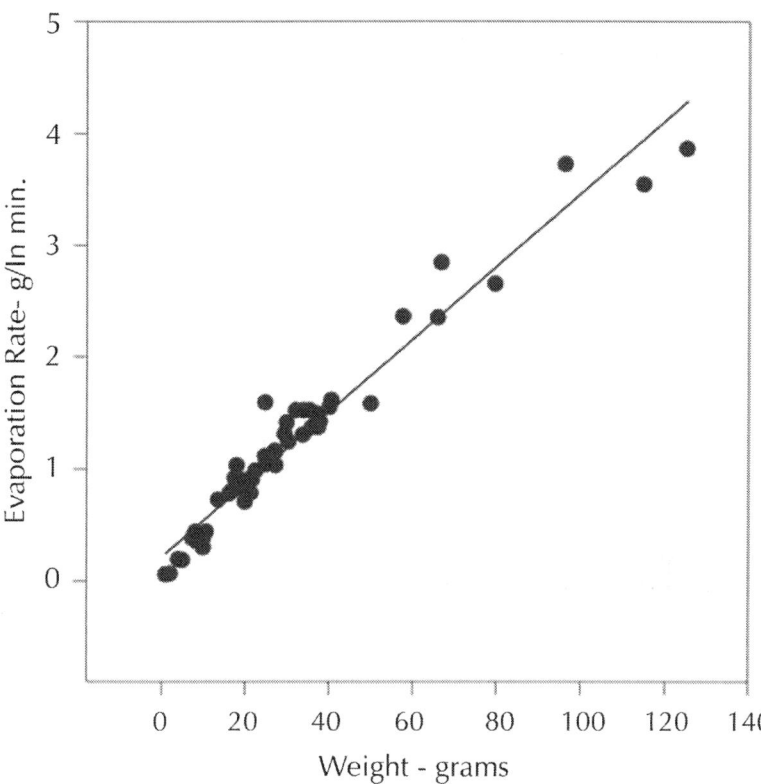

Figure 6: Correlation of mass with evaporation rate.

The evaporation rate data are illustrated in Figure 7. This figure shows that the evaporation rates of the pure hydrocarbons have a variable response to wind. Heptane (hydrocarbon number 7) shows a large difference between evaporation rate in wind and no wind conditions, indicating boundary-layer regulation. Decane (carbon number 10) shows a lesser effect and Hexadecane (carbon number 16) shows a negligible difference between the two experimental conditions. This experiment shows the extent of boundary-regulation and the reason for the small or negligible degree of boundary regulation shown by crude oils and petroleum products. Crude oil contains very little material with carbon numbers less than decane, often less than 3% of its composition [9]. Even the more volatile petroleum products, gasoline and diesel fuel only have limited amounts of compounds more volatile than decane, and thus are also not strongly boundary-layer regulated.

Saturation Concentration

Another evaluation of evaporation regulation is that of saturation concentration, the maximum concentration soluble in air. Table 4 lists the saturation concentrations of water and several oil components [10]. This table shows that saturation concentration of water is less than that of many common oil components. The saturation concentration of water is in fact, about two orders less in magnitude than the saturation concentration of volatile oil components such as pentane. This further explains why oil has a air boundary-layer limitation much higher than that of water and thus is not air boundary-layer regulated.

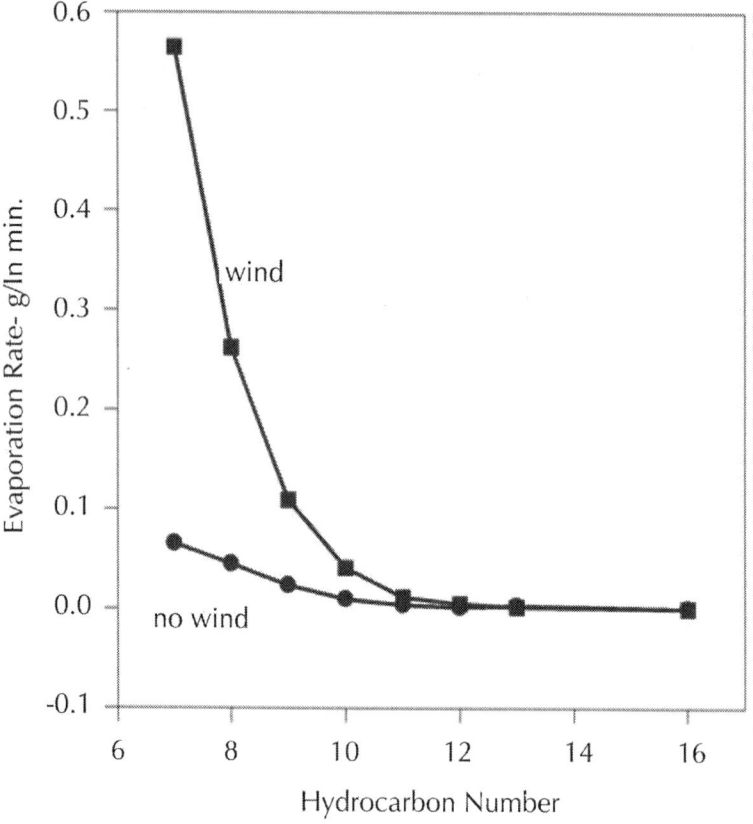

Figure 7: evaporation rates of pure compounds.

Table 4: Saturation concentration of water and hydrocarbons

Substance	Saturation Concentration* in g/m³ at 25°C
water	20
n-pentane	1689
hexane	564
cyclohexane	357
benzene	319
n-heptane	196
methylcyclohexane	192
toluene	110
ethybenzene	40
p-xylene	38
m-xylene	35
o-xylene	29

*Values taken from Ullman's Encyclopedia [10].

CONCLUSIONS

Oil evaporation is not air boundary-layer regulated. The results of the following experimental series have shown the lack of boundary-layer regulation: 1) a study of the evaporation rate of several oils with increasing wind speed shows that the evaporation rate does not change measurably with wind level. Water, known to be boundary-layer regulated, does show a significant increase with wind speed, U (U^x, where x varies from 0.5 to 0.78, depending on the turbulence level); 2) the volume or mass of oil evaporating correlates with the evaporation rate. This is a strong indicator of the lack of boundary-layer regulation because with water, volume (rather than area) and rate do not correlate; 3) evaporation of pure hydrocarbons with and without wind (turbulence) shows that compounds larger than nonane and decane are not boundary-layer regulated. Most oil and hydrocarbon products consist of compounds larger than these two and thus would not be expected to be boundary-layer regulated.

Having concluded that boundary-layer regulation is not specifically applicable to oil evaporation, it remains to explain why this is so. The reason is twofold: oil evaporation is relatively slow compared to the threshold where it would be air boundary-layer regulated; and the threshold to boundary-layer regulation for oil evaporation is much higher than that for water. These two factors were highlighted two ways:

1) A comparison of the maximum rates of evaporation for some oils, gasoline and water, in the absence of wind, shows that some oil rates exceed that for water by as much as an order of magnitude (water = 0.034 g/min, ASMB = 0.075 g/min, and Gasoline = 0.34 g/min; all under the specific conditions noted), and 2) The saturation concentration of several hydrocarbons in air reveals that some hydrocarbon saturation concentrations in air can be greater than that of water by as much as two orders-of-magnitude.

The fact that oil evaporation is not air boundary-layer regulated implies a simplistic evaporation equation will suffice to describe the process. The following factors do not require consideration: wind velocity, turbulence level, area, and scale size. The factors important to evaporation include time and temperature. Thickness is a factor above certain thicknesses, which are probably not relevant to a rapidly spreading oil slick. The latter is the subject of further experimentation.

REFERENCES

1. M. Fingas, "Oil and Petroleum Evaporation," Proceedings of the 34th Arctic and Marine Oilspill Program Technical Seminar, Vancouver, 4-6 October 2011, pp. 426- 459.
2. M. Fingas, "A Literature Review of the Physics and Predictive Modelling of Oil Spill Evaporation," Journal of Hazardous Materials, Vol. 42, No. 2, 1995, pp. 157-175.doi:10.1016/0304-3894(95)00013-K
3. W. Brutsaert, "Evaporation into the Atmosphere," Reidel Publishing Company, Dordrecht, 1982.
4. F. E. Jones, "Evaporation of Water," Lewis Publishers, Chelsea, 1992.

5. J. L. Monteith and M. H. Unsworth, "Principles of Environmental Physics," Hodder and Stoughton, London, 2008.
6. O. G. Sutton, "Wind Structure and Evaporation in a Turbulent Atmosphere," Proceedings of the Royal Society of London, Vol. 146, No. 858, 1934, pp. 701-722.
7. D. Mackay and R. S. Matsugu, "Evaporation Rates of Liquid Hydrocarbon Spills on Land and Water," The Canadian Journal of Chemical Engineering, Vol. 51, No. 4, 1973, pp. 434-439. doi:10.1002/cjce.5450510407
8. W. Stiver and D. Mackay, "Evaporation Rate of Spills of Hydrocarbons and Petroleum Mixtures," Environmental Science and Technology, Vol. 18, No. 11, 1984, pp. 834-840. doi:10.1021/es00129a006
9. Environment Canada, "Online Catalogue of Crude Oil and Oil Product Properties," 2011. http://www.etc-cte.ec.gc.ca/databases/OilProperties/oil_prop_e.html
10. Z. Wang and M. Fingas, "Oil and Petroleum Product Fingerprinting Analysis by Gas Chromatographic Techniques," In: L. M. L. Nollet, Ed., Chromatographic Analysis of the Environment, Taylor and Francis, Boca Raton, 2005, pp. 1027-1101.
11. "Ullmann's Encyclopedia," Ullmann Publishing, Hamburg, 2005-2009.

Chapter 8

Numerical Study of Oil/Water Separation by Ceramic Membranes in the Presence of Turbulent Flow

Tássia Mota Vieira[1], Josedite Saraiva de Souza[1], Enivaldo Santos Barbosa[1], Acto de Lima Cunha[1], Severino Rodrigues de Farias Neto[1], and Antonio Gilson Barbosa de Lima[2]

[1]Department of Chemical Engineering, Center of Sciences and Technology, Federal University of Campina Grande (UFCG), Campina Grande, Brazil

[2]Department of Mechanical Engineering, Center of Sciences and Technology, Federal University of Campina Grande (UFCG), Campina Grande, Brazil

ABSTRACT

Disposal of produced water during petroleum extraction causes serious environmental damage, hence the need to complete the water treatment before being disposed to environment within the criteria set established by environmental agencies in the countries. Ceramics membranes have been highlighted as a good device for separating oil/water. These act as a barrier to oil in the aqueous stream, because their essential properties for filtration, such as chemical inertness, biological stability and resistance to high temperatures. The limitation of the separation process is the decay of permeate flux during operation, due to concentration polarization and fouling. In this sense, this paper aims to evaluate numerically the feasibility of the process of separating oil/water by means of ceramic membranes in the presence of a turbulent flow induced by a tangential inlet. The results of the velocity, pressure and volumetric fraction distributions for the simulations different by varying the mass flow rate inlet and different geometric characteristics of the membrane are presented and analyzed.

INTRODUCTION

Knowledge of the process of separating water-oil is of great importance in the chemical, petrochemical, and food Industries, especially in solving problems related to environmental protection. Based on United States Environment Protection Agency (USEPA) regulations, the daily maximum limit for oil and grease is 42 mg/L and the monthly average limit is 29 mg/L. The Convention for the Protection of the Marine Environment of the North-East Atlantic (OSPAR Convention), the annual average limit for discharge of dispersed oil for produced water into the sea is 40 mg/L [1-4]. Thus, membrane technology has been intensively investigated as an alternative technique for the separation of stable emulsions of hydrocarbons [5], due to essential filtration properties, such as chemical inertness, biological stability and resistance to high temperatures contributing in the process of removing oils in mixtures with water [6].

The membrane separates immiscible solids and solutes that are dissolved, acting as a selective barrier allowing the passage of certain components while preventing the passage of others. In the process of separating two streams are produced: the concentrate stream containing the contaminants initially present in the feed stream and permeated or purified fraction of liquid passing through the membrane.

As reported in [7] ceramic membranes are of great interest in separation processes, due to their higher chemical and thermal stability when compared to polymeric membranes. By using ceramic membranes filtration can occur at temperatures above 500°C and at pH 1 to 14 and can be cleaned with aggressive chemicals substances, organic solvents or hot water vapor reflux.

Several studies have been reported in the literature using ceramic membranes as device to separate water/oil mixture [8-11] and have shown a separation efficiency ranging between 95% to 99.9% depending on the membrane and of the fluids properties that you want to separate However, these devices have shown a deficiency with respect to the permeate flow that is reduced over time due to fouling pores.

Given the above, this paper aims to present the numerical study of the behavior of fluids in the process of separating oil/water separation process through ceramic membrane in the presence of a turbulent flow induced by a tangential inlet. For the simulations we used the ANSYS CFX$^\rightarrow$ commercial software (code based on the CFD modeling-Computational Fluid Dynamics) which allows the simulation of systems involving fluid flow, heat transfer and other related physical processes.

PROBLEM DESCRIPTION

The study area corresponds to a cylindrical ceramic membrane that acts as a barrier to passage of particles oil droplet through the porous medium. This retention can be fully or partially restricting the carriage of one or more chemical species present in phases depending on the size of the molecules of the involved species. The fluid is injected tangentially in the membrane through a device located in one of its ends and flows along the membrane while the permeated is transported perpendicularly. Representation of the ceramic membrane can be observed in Figure 1, and its dimensions are presented in Table 1.

Two membranes configuration were studied: tubular (Figure 2) Grid with 446485 hexahedral elements and annular (Figure 3) grid with 197088 hexahedral elements. The annular space present in Figure 3 is the difference between the two geometries presented in this work.

MATHEMATICAL MODEL

Mathematical modeling is a physical representation of reality in the form of a set of consistent equations. The proposed mathematical model to describe the flow in porous media corresponds to a generalization of the equations conservation mass and momentum (NavierStokes) and Darcy's law applied to the Eulerian-Eulerian model of interfacial transfer and k-ε turbulence model (RNG). The renormalization group (RNG) k-ε model is similar in form to the k-ε model but includes additional terms for turbulence dissipation rate ε, furnishing more accurate predictions of the flow situations, including, separation process, streamlines, curves and stagnant regions, by considering the following simplifications:

- Newtonian incompressible fluid with constant physical and chemical properties;
- Stationary and isothermal flow;
- Mass transfer and interfacial momentum and mass source are neglected;
- The interfacial forces of non-drag force (forces of lift, wall lubrication, virtual mass, pressure and turbulent dispersion of solid) are neglected;
- The membrane walls of the tubular and annular geometries are statics.

With these considerations the resulting equations are: given as follows.

Numerical Study of Oil/Water Separation by Ceramic Membranes... 197

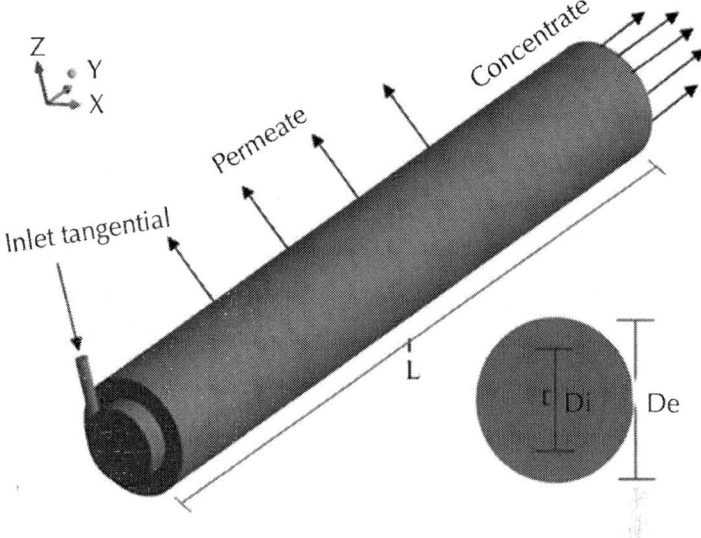

Figure 1: Geometrical representation of the ceramic membrane.

Table 1: Dimensions of the ceramic membrane

Length, L (cm)	67.8
Tangential inlet diameter (cm)	1.0
External diameter, De (cm)	10.0
Internal diameter, Di (cm)	6.5

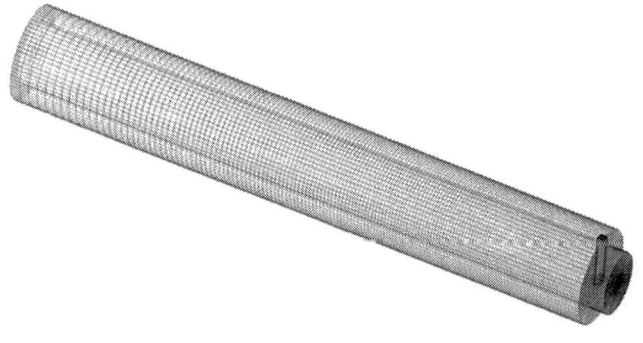

(a)

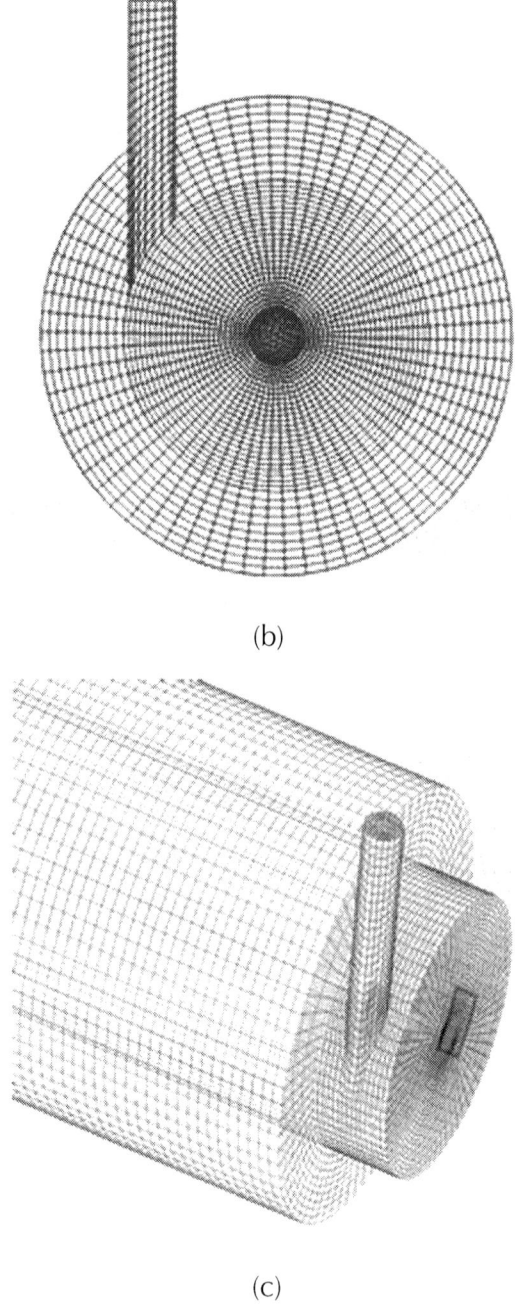

(b)

(c)

Figure 2: Representation of mesh with tubular ceramic membrane.

Numerical Study of Oil/Water Separation by Ceramic Membranes... 199

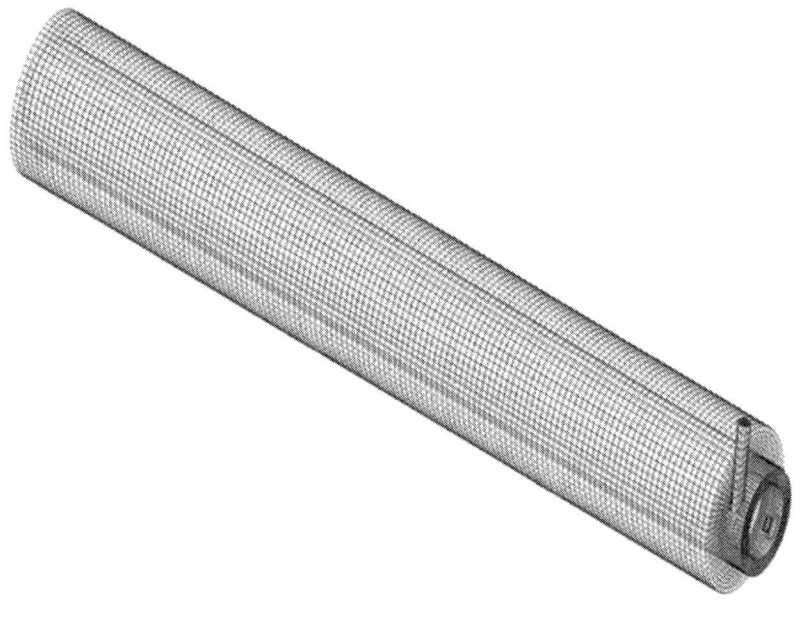

(a)

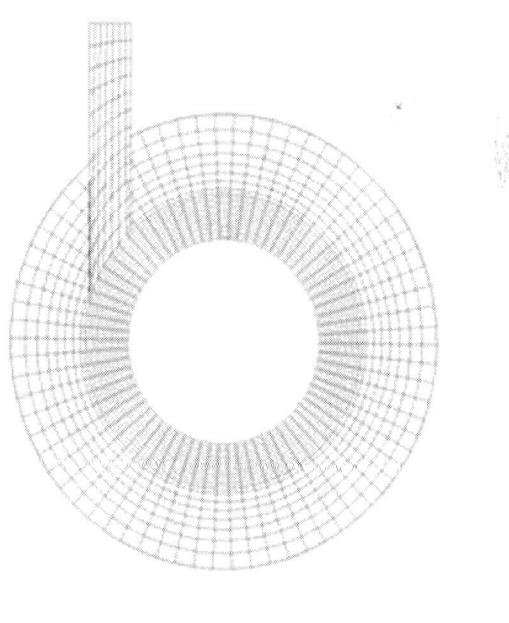

(b)

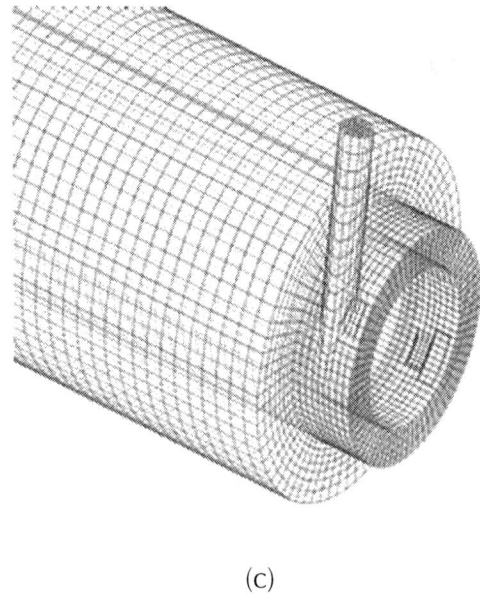

(c)

Figure 3: Representation of structured mesh with the annular geometry of the ceramic membrane.

Mass Conservation Equation for the Fluid Phase

The mass conservation is given by (1)

$$\nabla \cdot \left(f_\alpha \rho_\alpha U_\alpha \right) = 0 \tag{1}$$

Where the Greek sub-indices α represent the involved phases of water/ oil mixture; f, ρ, and $\vec{U}$ are respectively the volume fraction, density and vector velocity. For phase α, the vector velocity is given by $U_\alpha = (u, v, w)$.

Equation of Mass Conservation for the Medium Porous

The mass conservation equation for the flow in porous media is defined by the following equation:

$$\frac{\partial}{\partial t}(\phi\rho) + \nabla \cdot (\rho K U) = 0 \quad (2)$$

Where t is the time, f is the volumetric porosity and $K = (K^{ij})$ is the second order symmetric tensor called the tensor of porosity of the area.

Momentum Conservation Equation for the Fluids Phases

The linear momentum conservation for multiphase flow is defined by (3):

$$\nabla \cdot \left[f_\alpha (\rho_\alpha U_\alpha \otimes U_\alpha) \right] = -f_\alpha \nabla p_\alpha + \nabla \cdot \left\{ f_\alpha \mu_\alpha \left[\nabla U_\alpha + (\nabla U_\alpha)^T \right] \right\} + S_{M\alpha} + M_\alpha \quad (3)$$

where p is the pressure, $S_{M\alpha}$ represent the external forces per unit volume of the phase β for phase α. M_α describes total force The total force on phase α due to interaction with other phase (M_α) is given by (4):

$$M_\alpha = \sum_{\alpha \neq \beta} M_{\alpha\beta} \quad (4)$$

Where,

$$M_{\alpha\beta} = C^{(d)}_{\alpha\beta}\left(U_\beta - U_\alpha\right) \quad (5)$$

Where $C^{(d)}_{\alpha\beta}$ corresponds to the dimensionless drag coefficient given by (7).

$$c^{(d)}_{\alpha\beta} = \frac{3}{4}\frac{C_D}{d_p} f_\beta \rho_\alpha \left|U_\beta - U_\alpha\right| \quad (6)$$

Where d_p is the particle diameter and C_D is the drag coefficient. In this work, C_D has been adopted to be equal to 0.44 [12] (for the turbulent and viscous regime).

Momentum Transfer Equation for the Porous Medium

The momentum conservation equation for porous media is defined by the equation (7).

$$\nabla \cdot \left[\rho\phi(KU)\otimes U\right] = -\nabla \cdot \left\{\mu_e K\left[\nabla U + (\nabla U)^T\right]\right\} + S^M_i \quad (7)$$

Where μ_e is the effective viscosity, and S^M_i represents the source of quantity momentum linear, in [8] the source of quantity momentum linear is represented by:

$$S^M_i = -C^{R1} U_i - C^{R2}\left|U\right|U_i + S^{spec}_i \quad (8)$$

Where C^{R1} is the linear resistance coefficient, C^{R2} is the quadratic resistance coefficient and S_i^{spec} represents other sources of linear momentum quantity, related to the species present and U and U_i are superficial velocities.

Therefore, the Darcy's law generalized becomes:

$$-\frac{\partial P}{\partial x_i} = \frac{\mu}{K} U_i + K_{loss} \rho |U| U_i \tag{9}$$

Where μ denotes dynamic viscosity, K_{loss} is the empirical coefficient of loss in [12], $K_{loss} = 0$ and μ/K cannot be zero.

Comparing equations (1) and (2) and using actual speeds instead the superficial velocity, the coefficients C^{R1} and C^{R2} are expressed by:

$$C^{R1} = \frac{\mu}{\phi K} \tag{10}$$

And

$$C^{R2} = \frac{K_{loss} \rho}{\phi^2} \tag{11}$$

Turbulence Model

Due to complexity of the turbulent fluid flow through the membrane we use k-ε model (RNG) to complete the mathematical formulation. The renormalization group (RNG) k-ε model is similar in form to the k-ε model but includes additional terms for turbulence dissipation rate ε, furnishing more accurate predictions of the flow situations, including, separation process, streamlines, curves and stagnant regions.

The values of turbulent kinetic energy, k, and turbulent dissipation rate, ε are directly obtained from the differential equations of transport as can be observed in (12) and (13):

$$\frac{\partial}{\partial t}(\rho k) + \nabla \cdot (\rho U k) = P_k - \rho \varepsilon + \nabla \cdot \left[\left(\mu + \frac{\mu_t}{\sigma_{kRNG}} \right) \nabla k \right] \quad (12)$$

$$\frac{\partial}{\partial t}(\rho \varepsilon) + \nabla \cdot (\rho U \varepsilon)$$
$$= \nabla \cdot \left[\left(\mu + \frac{\mu_t}{\sigma_{\varepsilon RNG}} \right) \nabla \varepsilon \right] + \frac{\varepsilon}{k} \left(C_{\varepsilon 1 RNG} P_k - C_{\varepsilon 2 RNG} \rho \varepsilon \right) \quad (13)$$

Where μ is the dynamic viscosity, ρ is the density and μ_t is the turbulent viscosity which is given by (14).

$$\mu_t = C_\mu \rho \frac{k^2}{\varepsilon} \quad (14)$$

Where C_μ is an empirical constant and the values of the constants are given by:

$$C_\mu = \sigma_{kNRG} = \sigma_{\varepsilon NRG} = 0.7179 \quad (15)$$

And

$$C_{\varepsilon 2 RNG} = 1.68 \quad (16)$$

$$C_{\varepsilon 1 RNG} = 1.42 - \frac{\eta\left(1 - \frac{\eta}{4.38}\right)}{1 + \eta^3 \beta_{RNG}} \qquad (17)$$

Where,

$$\eta = \sqrt{\frac{P_k}{\rho \varepsilon C_{\mu RNG}}} \qquad (18)$$

In this equation $C_{\mu RNG}$ is the RNG turbulence model constant equal to 0.085 [12], P_k is the turbulence production due to viscous and buoyancy forces or shear production of turbulence, which is modeled using (19):

$$P_k = \mu_t \nabla U \cdot (\nabla U + \nabla U)^T + P_{kb} \qquad (19)$$

The term P_{kb} is the production of buoyancy and is modelled by (20) as follows:

$$P_{kb} = -\frac{\mu_t}{\rho \sigma_p} g \nabla \rho \qquad (20)$$

Where $\sigma_\rho = 1$.

The mathematical model does not predict the phenomenon of retention of particles or molecules in a porous medium, but consider difficulty or resistance to passage of the phases (oil and water) in porous media.

Boundary Conditions and Properties of Fluids and Membrane

To complement the mathematical modeling were defined boundary conditions for the simulated cases (Table 2).

The properties of water, oil and porous media used in this work are shown in Tables 3 and 4.

Cases Studies

Table 5 summarizes the six studied cases. Were analyzed different simulations varying the inlet velocity as well as some geometric characteristics of the membrane as shown earlier in Figures 2 and 3.

The numerical experiments were performed on a Server Quad-Core Intel Dual Xeon Processor E5430 of 2.66 GHz with 8 GB of RAM available in the laboratories LPFI (Fluid Dynamic Imaging Research Laboratory) and LCTF (Thermal and Fluid Computational Laboratory) of Chemical Engineering and Mechanical Engineering Department, respectively, Federal University of Campina Grande, Brazil The simulation time ranging from 3 to 4 days/simulation.

Table 2: Boundary conditions for the studied cases

Boundary	Type	*Condition
Feeding	Inlet	Volumetric fraction of water = 0.9
		Volumetric fraction of oil = 0.1
		Velocity of the feeding = Different for each case
Filtered	Outlet	Static pressure = 99,000 Pa
Inner wall of the membrane	Wall	All components of the velocity are zero
Walls of the device	Wall	All components of the velocity are zero
Outlet	Outlet	Static pressure = 98,000 Pa

*Based in experimental data.

Table 3: Physical-chemical properties of fluids

Proprieties	Water	Oil
Density (kg/m^3)	997.700	868.7
Viscosity (cP)	0.89	76
Molecular weight (kg/kmol)	18.015	873
Droplet diameter (mm)	-	0.010

Table 4: Properties of porous medium used in the simulations

Proprieties		Source
Porosity (-)	0.4403	Cunha et al. [13]
Permeability (m^2)	2.29 × 10^{-10}	Cunha et al. [13]

Table 5: Case studies considering the constant properties

Case	Inlet velocity (m/s)	Geometry
01	20	Membrane tubular
02	25	
03	30	
04	20	Membrane with annular space
05	25	
06	30	

RESULTS AND DISCUSSION

The numerical results obtained in research by the meshes presented in Figures 2 and 3 are given by means of streamlines, vector field and volumetric fraction distribution, as well as a graph representing the behavior of the filtered mass flow rate of the water and oil mixture for different inlet conditions of the mixture.

In Figures 4 and 5 are illustrated the behavior of streamlines to the oil and water inside the tubular and annular geometries provided with a ceramic membrane for the different inlet velocities, corresponding to mixture mas flowrate of (1.498; 1.872 and 2.247 kg/s) respectively.

By observing carefully the Figures 4 and 5 we see the presence of two upward streams distinct of fluids in spiral near the tangential inlet that disappears as it departs from the tangential inlet. The decrease of the spiral motion is related to the reduction of angular momentum and the axial momentum to dominate the flow along the geometry. This behavior leads to the appearance of more orderly flow leading to the velocity profile close to parabolic (see vector field in Figure 6) observed in flow in tubes in laminar regime. On the other hand, the difference in the behavior of the currents of water and oil is attributed to density difference between the phases and the balance of drag, centrifugal and weight forces throughout the apparatus.

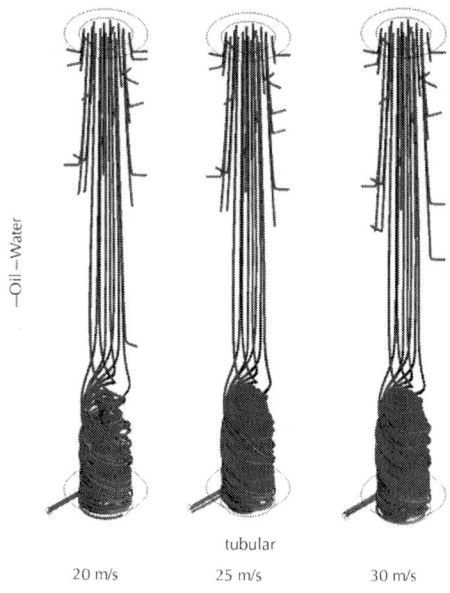

Figure 4: Streamlines in the tubular geometry, for differents inlet mixture velocity.

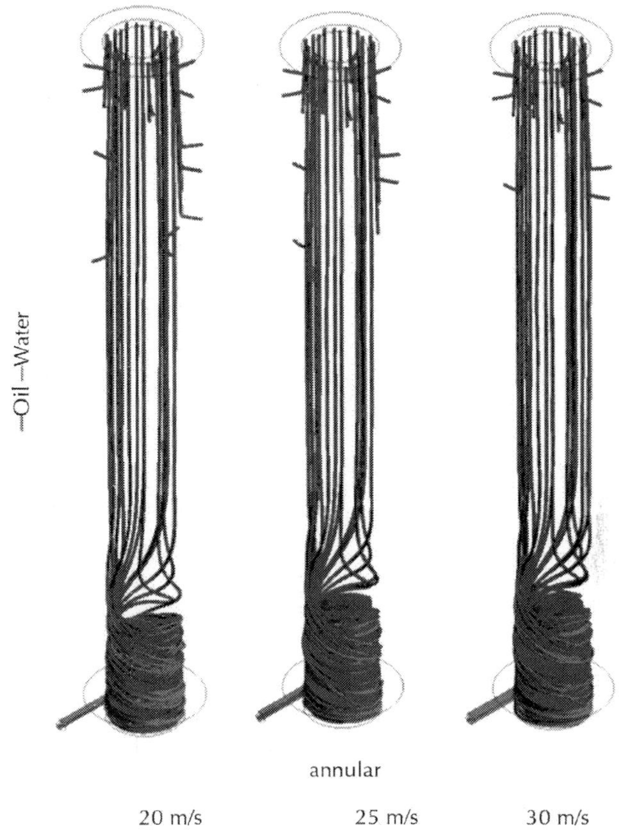

—Oil—Water

20 m/s   25 m/s   30 m/s

annular

Figure 5: Streamlines in the annular geometry, for differents inlet mixture velocity.

By observing Figure 7, where is represented a detail in the region near the tangential inlet for the two geometries analyzed (tubular and annular), one sees clearly the influence of the geometric aspect in the behavior of water and oil streamlines. The fact of insertion of a tube in the tubular device, forming an annular space, eliminates the region of mixing near the axial of the membrane.

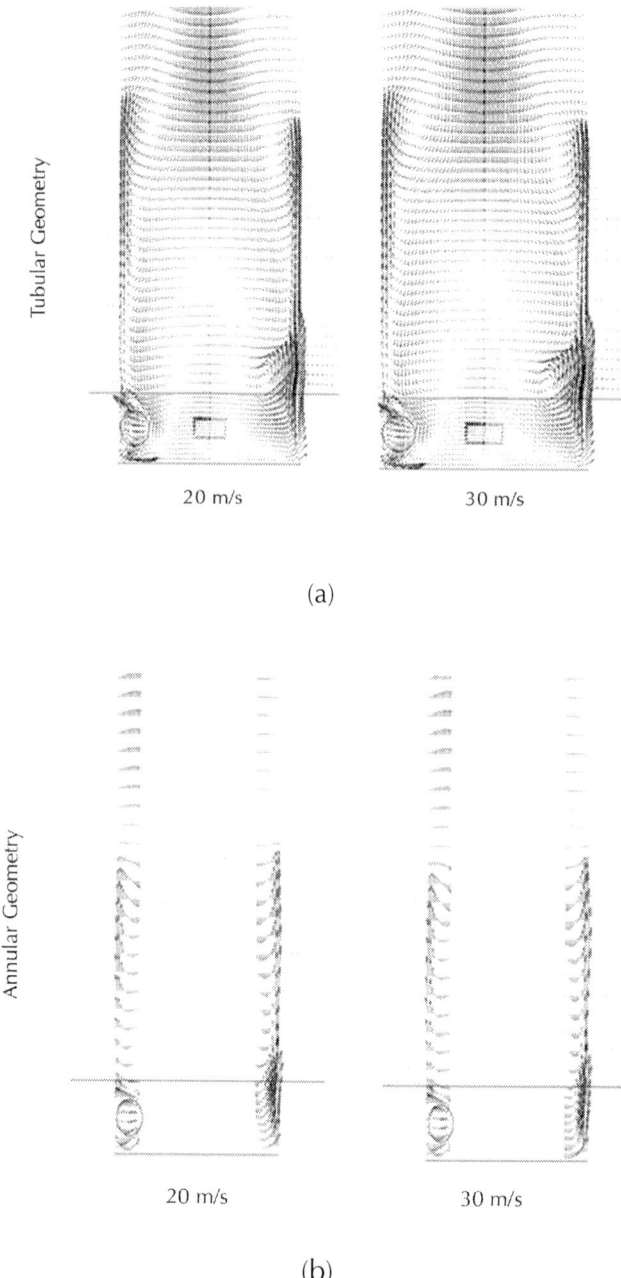

Figure 6: Details of the mixture velocty vector in (a) tubular and (b) annular geometries for different inlet mixture velocity.

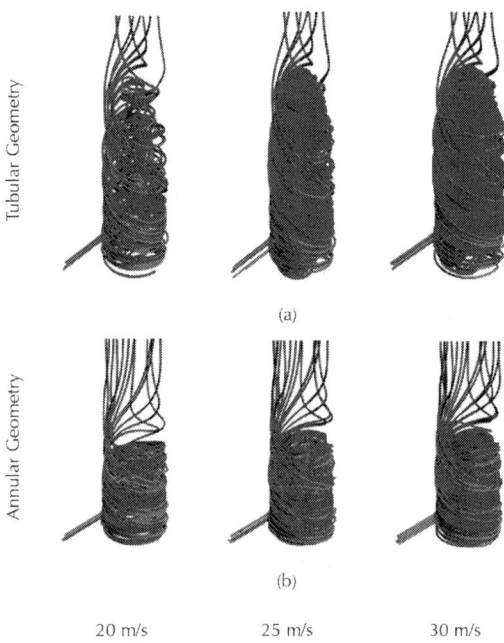

Figure 7: Details of the streamlines in (a) tubular and (b) annular geometries for different inlet mixture velocity.

This behavior is provided by reducing the pressure at the center of the tubular geometry (Figure 8). The results of the vector field of velocity shown in Figure 6. These results allows us to observe that the swirling motions (or recirculation zones) are more important, i.e., presents larger dimensions intensities, in tubular than in the annular geometry. This leads to a probable mixture of the water with the oil providing a greater dispersion of oil and the possible formation of emulsion between the phases.

Observing the behavior of the velocity vector field in the annular geometry it is clear that the length of the recirculation zones are smaller than those observed in the tubular geometry which provides a decrease in the swirling flow along the membrane starting from tangential inlet, from this position the axial momentum becomes dominant instead of the angular momentum.

Returning to Figure 8, where is represented the pressure distribution upon three plans around the tangential inlet, it is observed that the largest gradients are located in the porous medium, which ensures the permeation of fluid in the membrane in the radial direction thus providing the emergence of the filtrate. By observing carefully, one realizes that these gradients are most important to the annular geometry than for the tubular, which indicates a greater amount of filtrate volumetric and consequently a greater amount of oil present in the filtrated. This situation can be verified by observing the xy plane for the two geometries in Figure 9, which depicts the volumetric fractions distributions of oil on these plans. It is clearly shown that the annular geometry has more permeation of the oil into the membrane. This phenomenon is possibly a problem for the membrane because there may be an obstruction or greater drag of oil by water across the membrane.

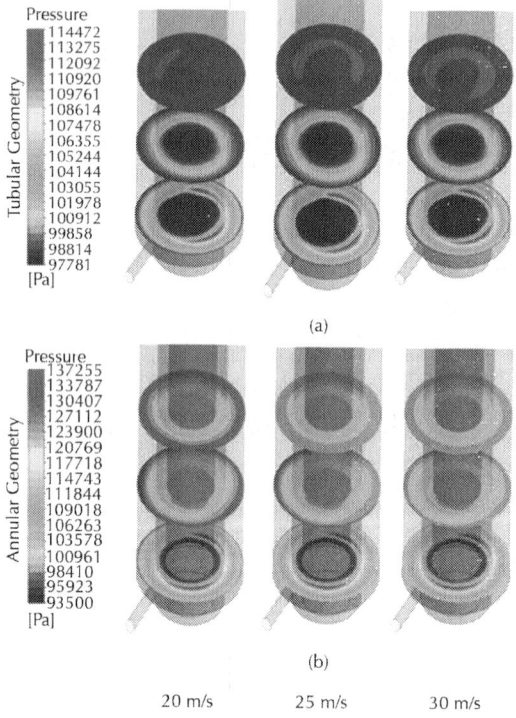

(a)

(b)

20 m/s   25 m/s   30 m/s

Figure 8: Representation of the pressure distribution on the xz planes for tubular and annular geometries for the differents inlet velocity.

The Figures 10 and 11 are represented the filtered water and oil mass flow rate as a function of mixture mass flow rate in the feed for tubular and annular geometries. These figures show that there is an increase of filtered water mass flow rate as well for oil, for the annular geometry compared to the tubular geometry. While there has been an increase in oil concentration in the filtrate using the annular geometry, it is clear that there is a very significant increase in the amount of filtered water.

It should be noted that even with the low increase in oil concentration in the filtrate using the annular geometry, there was an increase in the production of treated water compared to the tubular geometry. This high concentration of oil may be related to the permeability adopted in this work, so if it reduced the value of permeability probably will reduce the oil concentration in the filtrate.

CONCLUSIONS

From the numerical results of the oil/water separation process using membrane with tubular and annular geometries, under the evaluated conditions, we can be concluded that:

- The proposed mathematical model was able to predict the fluid-dynamics for the separation process of water and oil mixture by ceramic membrane;
- It was possible to observe a three-dimensional behavior of the flow within the two geometries evaluated;
- The fluid-dynamic behavior inside the geometries showed a decrease in turbulence intensity of fluids starting from the tangential inlet of the membrane;
- The increase in feed streams provided an increase in the filtered water and oil mass flow rate for the annular and tubular geometries
- The annular geometry presented higher values of filtrate for both the oil to water as compared to the tubular geometry. However, it requires a further examination of the parameters of the porous medium, such as permeability, to minimize the presence of oil in the filtrate.

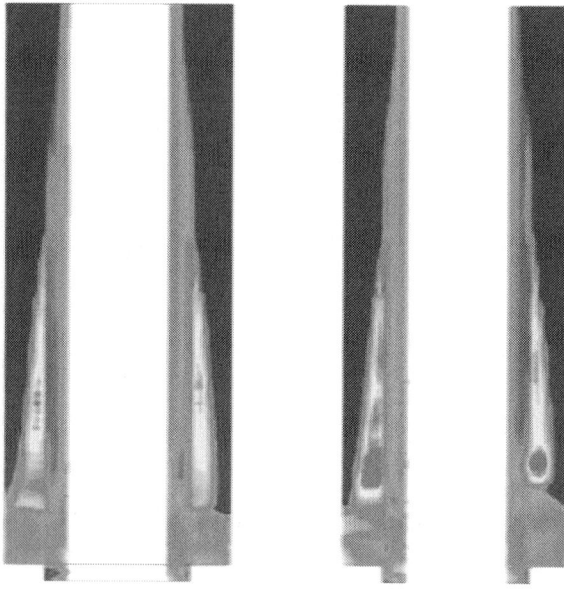

(a)

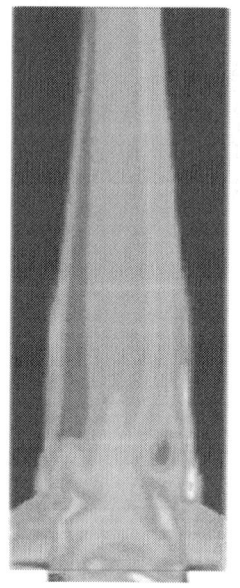 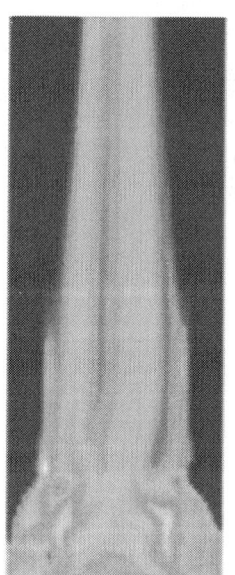

(b)

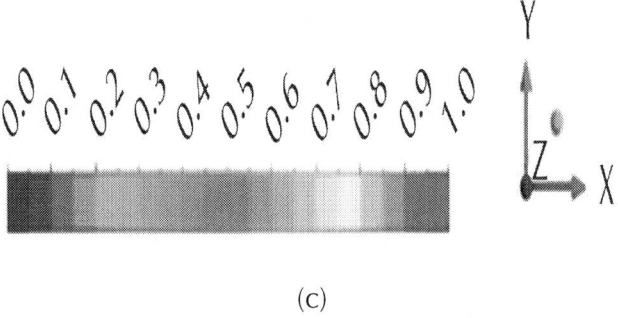

(c)

Figure 9: Volumetric fraction distribution on the xy plane for the tubular and annular geometries for the inlet mixturevelocity equal to 20 m/s.

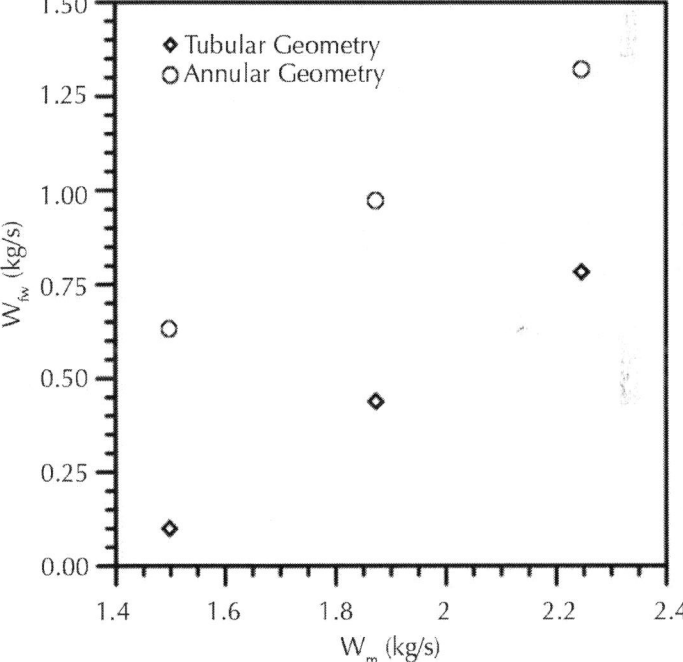

Figure 10: Filtrate water mass flow rate as a function of the inlet mixture mass flow rate for tubular and annular geometries.

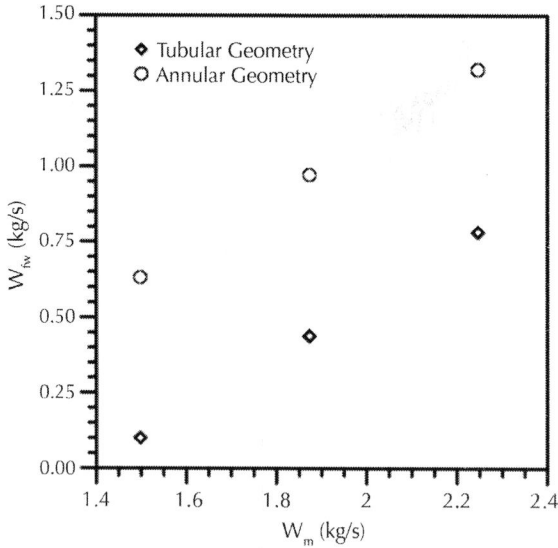

Figure 11: Filtrate oil mass flow rate as a function of the inlet mixture mass flow rate for tubular and annular geometries.

ACKNOWLEDGEMENTS

The authors would like to express their thanks to CNPQ, CAPES, FINEP and ANP. (Brazil), for supporting this work, and are also grateful to the authors cited in the text that helped in the improvement of quality.

REFERENCES

1. J. M. Neff, "Bioaccumulation in Marine Organisms: Effect of Contaminants from Oilwell Produced Water," Elsevier, Netherlands, 2002. http://www.epa.gov
2. OSPAR Commission, "Report on Discharges Spills and Emissions from Offshore Oil and Gas Installations," 2005. http://www.ospar.org/documents/dbase/publications/p00221 Offshore%20report%202003.pdf
3. F.-R. Ahmadun, et al., "Review of Technologies for Oil and Gas Produced Water Treatment," Journal of Hazardous

Materials, Vol. 170, No. 2-3, 2009, pp. 530-551. doi:10.1016/j.jhazmat.2009.05.044
4. M. A. A. Zaini, R. G. Holdich and I. W. Cumming, "Crossflow Microfiltration of Oil in Water Emulsion via Tubular Filters: Evaluation by Mathematical Models on Droplet Deformation and Filtration," Jurnal Teknologi, Vol. 53, 2010, pp. 19-28.
5. S. R. H. Abadi, M. R. Sebzari, M. Hemati, F. Rekabdar and T. Mohammadi, "Ceramic Membrane Performance in Microfiltration of Oily Wastewater," Desalination, Vol. 265, No. 1-3, 2011, pp. 222-228. doi:10.1016/j.desal.2010.07.055
6. S. Benfer, U. Popp, H. Richter, C. Siewert and G. Tomandl, "Development and Characterization of Nano-filtration Membranes," Separation and Purification Technology, Vol. 22-23, 2001, pp. 231-237. doi:10.1016/S1383-5866(00)00133-7
7. S. Ahmed, M. T. Seraji, J. Jahedi and M. A. Hashib, "CFD Simulation of Turbulence Promoters in a Tubular Membrane Channel," Desalination, Vol. 276, No. 1-3, 2011, pp. 191-198. doi:10.1016/j.desal.2011.03.045
8. S. R. Lautenschlager, S. S. F. Filho and O. Pereira, "Mathematical Modeling and Operational Optimization of Ultrafiltration Membrane Processes," Engenharia Sanitária Ambiental, Vol. 14, 2009, pp. 215-222. doi:10.1590/S1413-41522009000200009
9. D. F. Maia, "Development of Ceramic Membranes for Oil/Water Separation," Ph.D. Thesis, Federal University of Campina Grande, Campina Grande, 2006.
10. R. Del. Colle, R. N. Haneda, E. Longo, M. J. Godinho and S. R. Fontes, "Method of Chemical Impregnation Applied to Microporous Tubes and Tubular Membranes for the Microfiltration of Emulsions and Suspensions of Bacteria," Cerâmica, Vol. 54, 2008, pp. 21-28.
11. ANSYS CFX, "User Manual Theory," ANSYS CFX-Solver Theory Guide, ANSYS Inc., Canosburg, 2009.
12. A. L. Cunha, S. R. F. Neto and H. L. Lira, "Application of Fluid Dynamic Computational Oil/Water Separation in through Selective Membranes," Proceedings of the 3rd Congress of Scientific Initiation of the Federal University Campina Grande, Campina Grande, 2006, pp. 1-11.

Chapter 9

Modeling, Simulation and Optimization of Continuous Gas Lift Systems for Deepwater Offshore Petroleum Production

J.N.M. de Souza[a], J.L. de Medeiros[a, 1], A.L.H. Costa[b, 2], and G.C. Nunes[c, 3]

[a]School of Chemistry, Federal University of Rio de Janeiro, UFRJ, Centro de Tecnologia, Bloco E, sala 209, Cidade Universitária, Ilha do Fundão, CEP 21949-900, Rio de Janeiro, RJ, Brazil

[b]Institute of Chemistry, Rio de Janeiro State University, UERJ, Rua São Francisco Xavier, 524, Pavilhão Haroldo Lisboa da Cunha, Maracanã, CEP 20550-900, Rio de Janeiro, RJ, Brazil

[c]Petróleo Brasileiro S.A.- PETROBRAS, Av. República do Chile, 65, Centro, CEP 20031-912, Rio de Janeiro, RJ, Brazil

ABSTRACT

This paper proposed a framework for the analysis of continuous gas lift systems using an optimization algorithm coupled to a stationary two-phase flow network model. The objective function can consider the annualized capital costs on compressor, turbine and gas pipelines, the operating costs related to fuel and the revenue from produced oil. The interaction among wells, production lines and riser is properly evaluated by a stationary two-phase flow simulator for pipe networks composed by mass balances at network elements and momentum balances at pipes using the Beggs and Brill empirical correlation. The solution of the optimization problem can estimate important information for the conceptual design phase of a petroleum production system: (i) the injected gas flow rates that guarantees maximum oil production, (ii) the injected gas flow rates for maximum profit and (iii) optimal design of gas lift system considering capital costs of compressor, turbine and gas pipelines. Case studies of single and multiple wells with different complexities describe some applications of the proposed framework. At the first case study, an offshore petroleum well with gas lift artificial elevation is simulated – to determine the behavior of petroleum production as a function of the injected gas flow rate for different reservoir pressures and different wellbore diameters – and optimized—to determine the maximum production considering different productivity indexes. At the second case study, a complex petroleum production system with multiple wells is simulated and optimized to obtain the optimal design considering annualized costs of compressor, turbine driver, gas pipelines and fuel gas consumption.

INTRODUCTION

The high oil prices increase efforts in exploration and production, development of marginal fields and enhanced oil recovery projects. "Worldwide average oil recovery factor is expected to increase substantially from the current figure of around 35% due to technology development, adding significant resources to the reserve base" (Kjärstad and Johnsson, 2009).

In many cases of deepwater production, when the reservoir pressure is not sufficient to guarantee the oil elevation up to surface with a viable economical return, the necessity of artificial lift technologies to enhance the recovery factor is mandatory. A very common and efficient technique is the gas lift, where the injection of lean gas in a certain position of the well reduces the mean density of the liquid column and thus decreases the hydrostatic pressure. Two methodologies are commonly applied: Continuous Gas Lift (CGL) and Intermittent Gas Lift (IGL). As this work is restricted to stationary two-phase flow, only the CGL is considered.

The quantity of injected gas is a critical variable whereas a lower value can reduce significantly the production and a higher value can increase the operational costs with compression and gas usage. In most cases, it is possible to verify the oil production that reaches a maximum value for a certain injected gas flow rate.

Many authors explored this optimization problem determining the optimal operational conditions to extract the maximum quantity of oil for single well models (Fang and Lo, 1996) and multiple wells model (Alarcon et al., 2002 and Ray and Sarker, 2007) and considering or not the constraints on gas availability and using different formulations: linear programming (Fang and Lo, 1996), mixed integer linear programming (Kosmidis et al., 2005), non-linear programming (Alarcon et al., 2002), dynamic programming (Camponogara and Nakashima, 2006) and genetic algorithms (Ray and Sarker, 2007).

As described by Dutta-Roy and Kattapuram (1997), the gas lift optimization problem shall consider the effect of flow interactions between two wells sharing a common flow line. This is a serious limitation for the current major commercial reservoir simulators where the gas lift allocation is considered separately.

Excepting the works of Dutta-Roy and Kattapuram, 1997 and Kosmidis et al., 2005, the literature of gas lift optimization is restricted to the evaluation of each well isolated using a simplified hydrodynamic model. In other words, network effects commonly found in complex subsea systems, where two or more wells share a production line, are not considered.

The emphasis of this work is to propose an optimization problem coupled to a two-phase flow pipe network model to determine the gas allocation and the gas lift system design considering three types of objective functions: (i) maximization of produced oil, (ii) maximization of profit (difference of revenue from produced oil and costs with fuel gas) and (iii) design optimization of gas injection system.

The present work evaluates the complex behaviors of two-phase flow using a proper network model to determine stationary states of important variables like pressure, mass flow rates and hold-up along the wells and pipelines required by the objective function. The model specifications are the injected gas mass flow rate, the produced oil flow rate and the pressure at the riser outlet. The outputs of interest are the pressure at the lower height of the well and the pressure at the gas compressor discharge.

TWO-PHASE FLOW MODELING

The utilization of pipe network is the most economical and safe alternative for fluid transportation. The modern urban and industrial infrastructure is based on transportation of large amounts of liquids and gases, for instance, oil pipelines connecting production centers and refineries, multiproduct pipe networks from petrochemical plants to customers, potable water distribution networks to urban centers and many other applications.

This paper is focused on a two-phase flow pipe network, a device with the objective of transporting gas and liquid simultaneously. A very complex example of these networks is present at the offshore petroleum production systems, where gas, oil and water flow from multiple wells to the production line, the riser and finally to the production platform. The problem of gas and liquid simultaneous flows had been extensively studied in the past decades accompanying a vertiginous increasing demand of engineering solutions for progressively more and more difficult technological challenges.

The calculation of pressure gradient in single pipes is a consolidated theme, but a repetitive and time demanding series of steps is required when the engineer team needs to determine the pressure along a set of interconnected pipes. It is important to highlight that many authors focused on two-phase flow for single pipe application causing a lack of works describing application of simplified models to network context. An exception is the work of Floquet et al. (2009) that presented a network model able to perform multiphase simulations from the wellbore to the surface facilities with each pipeline calculated via commercial software and an interconnection algorithm to guarantee the mass and momentum closure at each line extremity. Abel Waly et al. (1996) presented a comparison between a network model and measurements from Zeit Bay field, but the modeling approach and the simulation environment are not detailed in this work.

With this motivation, a two-phase flow model for gas–liquid pipe networks is presented based on Beggs and Brill correlation (Beggs and Brill, 1973)—a practical and classical two-phase flow correlation that is able to determine pressure gradient and hold-up in pipes with any inclination and under any flow pattern. Some simplifications guarantee a robust model that is mathematically represented by a system of non-linear algebraic equations (coupled with ordinary differential equations) and solved with the Newton–Raphson method. The initial estimation for the iterative solver is generated by a linear model considering laminar flow without momentum exchange between phases.

Momentum Balance

Let the control volume presented in Fig. 1, which is the stationary two-phase flow model based on a single momentum balance of the mixture be written as Eq. (1):

$$A\frac{dP}{dx} + q_L\frac{dv_L}{dx} + q_G\frac{dv_G}{dx} + A\rho_s g \sin(\theta) + \pi D \tau_m \left(1 + \frac{L_{eq}}{L}\right) = 0 \quad (1)$$

where D is the pipe diameter in m, A is the cross section area of the pipe in m², L is the pipe length in m, L_{eq} is the pipe equivalent length of accidents and fittings in m, P is the mean pressure of the pipe in Pa, q_L and q_G are the mass flow rate in kg/s of liquid and gas, respectively, v_L and v_G are velocities in m/s of liquid and gas, respectively, θ is the inclination angle in rad, ρs is the mixture density in kg/m³ and τm is the mixture shear stress in Pa.

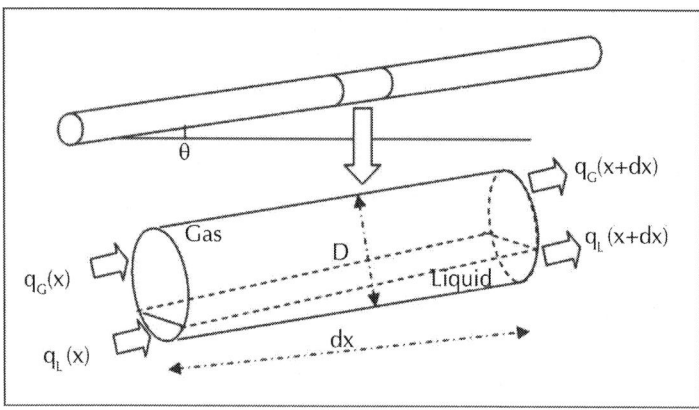

Figure 1: Control volume for momentum balance.

Rearranging Eq. (1), it is possible to evaluate the pressure drop with the following equation:

$$\frac{dP}{dx} = -\frac{A\rho_s g \sin(\theta) + \pi D\tau_m \left(1 + \frac{L_{eq}}{L}\right)}{(A + a_L q_L + a_G q_G)} \quad (2)$$

The terms a_L and a_G are coefficients related to the gas compressibility:

$$a_L = \left(\frac{D_P v_{sG}}{\rho_G}\right)\left(\frac{D_{vsG} v_L}{H_L}\right) \quad (3)$$

$$a_G = -\left(\frac{D_P v_{sG}}{\rho_G}\right)\left(\frac{D_{vsG} v_G}{H_G} + 1\right) \quad (4)$$

where H_L and H_G are hold-ups in terms of area fraction of liquid and gas, respectively, v_{sL} and v_{sG} are, respectively, liquid and gas superficial velocities in m/s, DP is the derivative of gas density $_G$ in respect to pressure P and D_{vsG} is the derivative of liquid hold-up in respect to gas superficial velocity. For simplification, in this work D_{vsG} is numerically calculated.

The closure relation that permits the calculation of m'_s and D_{vsG} composed only by phenomenological equations and valid for a large range of flow rates is an open task most due to the phenomenon complexities regarding the interaction between phases. Some authors presented mechanistic models valid for restricted conditions but used certain phenomenological aspects, as for instance, Aziz et al., 1972, Taitel and Dukler, 1976, Oliemans, 1987, Taitel and Barnea, 1990, Hasan and Kabir, 1992 and Ansari et al., 1994. Other group of authors developed two-phase flow correlations valid for different conditions based on pilot plant facilities: Lockhart and Martinelli, 1949, Flanigan, 1958, Hughmark, 1962, Hagedorn and Brown, 1965, Eaton et al., 1967, Orkiszewski, 1967, Beggs and Brill, 1973, Oliemans, 1976 and Mukherjee and Brill, 1985.

For this paper, the Beggs and Brill (1973) correlation is chosen because it is one of the few published correlations capable of describing with appropriate accuracy the behavior of two-phase flow for all flow patterns and for any pipe inclination (Fig. 2).

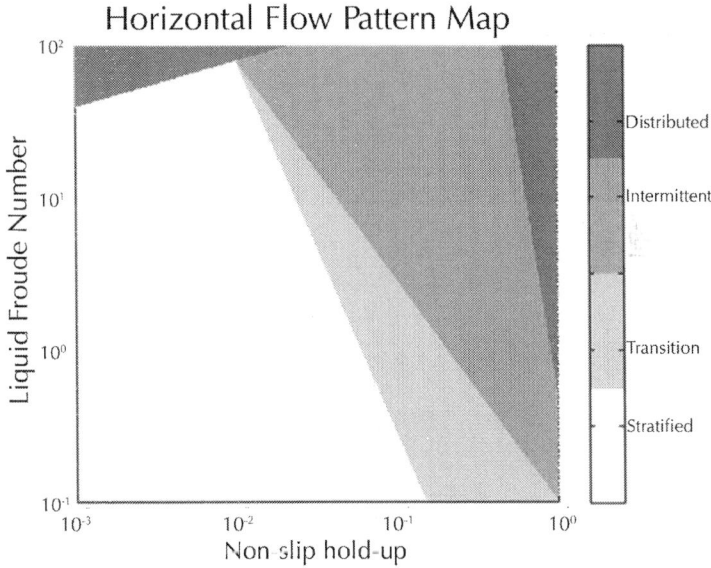

Figure 2: Beggs and Brill flow map.

TWO-PHASE FLOW PIPE NETWORK

Definitions

Pipe networks are constituted by a set of elements as pipelines, valves, pumps, compressors and other devices interconnected to promote fluid transport between sources and client points.

A systematic methodology shall be implemented in order to properly organize the model due to the complexity of element interconnection. A very common approach widely used for network simulation is based on direct graphs or digraphs. This work uses the same notation presented in Mah, 1990 and Costa et al., 1998, where a network is composed by N nodes and S lines.

The fluid flow along a pipe network can be represented by the following set of $3N + 4S$ variables:

(1) Nodal variables

(1.1) Vector of external mass flow rates of liquid w_L (dimension $N \times 1$, where $w_L i$ is the external mass flow rate of liquid at the i-th node, in kg/s).

(1.2) Vector of external mass flow rates of gas w_G (dimension $N \times 1$, where $w_G i$ is the external mass flow rate of gas at the i-th node, in kg/s).

(1.3) Vector of nodal pressures PW (dimension $N \times 1$, where PWi is the stagnant pressure at the i-th node, in bar).

(2) Line variables

(2.1) Vector of internal mass flow rates of liquid q_L (dimension $S \times 1$, where $q_L j$ is the internal mass flow rate of liquid at the j-th line, in kg/s).

(2.2) Vector of internal mass flow rates of gas q_G (dimension $S \times 1$, where $q_G j$ is the internal mass flow rate of gas at the j-th line, in kg/s).

(2.3) Vector of pressures at pipe inlets P_I (dimension S × 1, where P_Ij is the pressure at the j-th pipe inlet extremity, in bar).

(2.4) Vector of pressures at pipe outlets P_O (dimension S × 1, where P_Oj is the pressure at the j-th pipe outlet extremity, in bar).

For the internal mass flow rate variables, a positive value indicates that the flow direction is the same of the digraph orientation and negative value indicates that flow is at an opposite direction in relation to the digraph orientation. For external mass flow rates, positive value indicates that fluid is entering the network and a negative value indicates that fluid is leaving the network.

It can be considered as four different applications for the nodes: clients (negative external mass flow rate), sources (positive external mass flow rate), connectors (null value of external flow rate) and fixed pressure (fixed value for stagnant pressure).

Matrix Notation

An important tool for the organization of the mathematical models is the digraph matrix notation based on incidence matrices M, M_1 and M_2 with N lines and S columns, with the following generation rules:

- M has the element of i-th row and j-th column equal to + 1 if the j-th line is connected to the i-th node and the node is at the end of the j-th line.
- M has the element of i-th row and j-th column equal to − 1 if the j-th line is connected to the i-th node and the node is at the beginning of the j-th line.
- M has the element of i-th row and j-th column equal to 0 if the j-th line is not connected to the i-th node.
- M_1 has elements equal to one at the positions were M is equal to one and zero for the other positions
- M_2 has elements equal to one at the positions were M is equal to minus one and zero for the other positions.

To organize the network variables, a column vector y is made via concatenation of all variable vectors as:

$$y = \begin{bmatrix} q_L^T & q_G^T & P_I^T & P_O^T & P_W^T & w_L^T & w_G^T \end{bmatrix}^T \qquad (5)$$

The model specifications are set according the specification vector e containing the values of the specified variables and the specification matrix E, where the element of i-th row and j-th column is equal to one if the j-th variable is specified at the i-th equation, otherwise is zero.

The pipe network model requires N + S specifications. In order to guarantee a feasible solution for the proposed model, only nodal variables are specified (for instance, pressures and/or external mass flow rates) and at least one pressure specification shall be provided.

Premises are adopted to guarantee a simplified representation of the two-phase flow phenomena without the necessity of thermal energy balances: (i) liquid phase is incompressible and (ii) gas phase with density described by a polytrophic relation. For isothermal flow the gas polytrophic coefficient shall be equal to one. For non-isothermal flows the polytrophic coefficients shall be estimated according to each case.

The parameters of the models are: gravitational acceleration g (m/s^2), liquid and gas density at reference pressure ρ_{L0} and ρ_{G0} (kg/m^3), liquid and gas polytrophic coefficient γ_L and γ_G, reference pressure P_0 (bar), liquid and gas viscosity μ_L and μ_G (kg/(m s)), interfacial tension (Pa m), incidence matrix M, specification matrix E, specification vector e, nodal elevation vector z (m), pipe inclination vector θ (rad), pipe diameter vector D (m), pipe length vector L (m), pipe equivalent length vector L_{eq} (m) and pipe roughness vector ε(m), with , D and as functions of the axial position along the pipe.

The complexity of two-phase flow implies a non-linear model presented in Section 3.3 that is solved using an iterative method through a feasible initial estimate obtained with a simplified linear model, which is presented in Section 3.4.

Non-Linear Model

The model equations are divided in four classes:
(1)   Nodal mass balances (2N equations)

$$\underline{M}\,\underline{q_L} + \underline{w_L} = \underline{0}$$

(6)

$$\underline{\underline{M}}\underline{q_G} + \underline{w_G} = \underline{0} \tag{7}$$

(2) Momentum balance at pipes, resultant from the integration of Eq. (2) along the pipe length (Sequations):

$$\underline{P_O} - \underline{P_I} - 0.5\left(\underline{\Delta P_1} + \underline{\Delta P_2}\right) = \underline{0} \tag{8}$$

with:

$$\Delta P_{1,i} = 10^{-5} \int_0^{L_i} \frac{dP_i}{dx} dx \tag{9}$$

$$\Delta P_{2,i} = 10^{-5} \int_{L_i}^0 \frac{dP_i}{dx} dx \tag{10}$$

(3) Momentum balance at nodes (2S equations):

$$\underline{\underline{M_1}}^T \underline{P_W} - \left(\underline{P_I} + \underline{h_{LI}} \times \underline{q_L} + \underline{h_{GI}} \times \underline{q_G}\right) = \underline{0} \tag{11}$$

$$\underline{\underline{M_2}}^T \underline{P_W} - \left(\underline{P_O} + \underline{h_{LO}} \times \underline{q_L} + \underline{h_{GO}} \times \underline{q_G}\right) = \underline{0} \tag{12}$$

with:

$$\underline{h_{ij}}_{\substack{i \in \{L,G\} \\ j \in \{I,O\}}} = 10^{-5} \text{diag}^{-1}\left(2\underline{\rho_{ij}} \times \underline{\rho_{ij}} \times \underline{A_j}\right)|\underline{v_{ij}}| \tag{13}$$

(4) Nodal specification equations (N + S equations)

$$\underline{\underline{E}}\,\underline{y} - \underline{e} = \underline{0} \tag{14}$$

where h_{LI} and h_{GI} are related to the kinetic energy of the liquid and gas at the line inlet, h_{LO} and h_{GO} are related to the kinetic energy of the liquid and gas at the line outlet, A_O and A_I are the cross section area of outlet and inlet extremities of pipes, v_{LO} and v_{GO} correspond to liquid and gas velocities at the outlet extremity of pipes and v_{LI} and v_{GI} correspond to liquid and gas velocities at the inlet extremity of pipes.

It was verified that an increase in the performance of the numerical methods when the pressure drop along the pipe is obtained as a mean value of pressure drop integrated according to different orientations as represented in Eq. (8). This approach guarantees that the momentum balance at a pipe has a similar sensitivity for changes at outlet pressure and inlet pressure. The determination of $P1,i$ according to Eq. (9) requires the integration of Eq. (2) from the inlet to the outlet considering an initial value for pressure equal to Pli. The determination of $P2,i$ according to Eq. (10) requires the integration of Eq. (2) from the outlet to the inlet considering an initial value for pressure equal to POi.

It is important to highlight that this model can only be applied to non-cyclic networks and without divergent nodes (where two or more lines leave the same node). This limitation is due the simplified nodal model adopted in this paper.

Linear Model

The objective of this linear model is to generate a feasible initial estimative for the iterative method. This model considers that: (i) one phase flows without the interference of the other phase, (ii) phases are incompressible and (iii) phase hold-ups H_L are known. With these considerations, it is possible to calculate the total pressure gradient as a linear combination of each phase. Additionally, laminar flow of both phases is considered, generating a linear system of equations. Thus, the momentum balance at pipes in Eq. (8) is substituted by Eq. (15):

$$\underline{P_I} - \underline{P_O} - \underline{\underline{F_L}}\,\underline{q_L} - \underline{\underline{F_G}}\,\underline{q_G} - \underline{d} = \underline{0} \tag{15}$$

where:

$$\underline{\underline{F_L}} = \frac{1.28 \cdot 10^{-3} \mu_L}{\pi \rho_L} diag\left(\underline{H_L}\right) diag(\underline{L}) diag^{-1}(\underline{D}) \tag{16}$$

$$\underline{\underline{F_G}} = \frac{1.28 \cdot 10^{-3} \mu_G}{\pi \rho_G} diag\left(\underline{1-H_L}\right) diag(\underline{L}) diag^{-1}(\underline{D}) \tag{17}$$

$$\underline{d} = 10^{-5} g \left(\rho_L diag\left(\underline{H_L}\right) + \rho_G diag\left(\underline{1-H_L}\right)\right) \underline{\underline{M}}^T \underline{z} \tag{18}$$

Disregarding the velocity terms of the momentum balance of Eq. (13), the momentum balances at nodes, Eqs. (11) and (12), are substituted by Eqs. (19) and (20).

$$\underline{\underline{M_1}}^T \underline{P_W} - \underline{P_1} = \underline{0} \tag{19}$$

$$\underline{\underline{M_2}}^T \underline{P_W} - \underline{P_0} = \underline{0} \tag{20}$$

This simplified model can be expressed using the matrix notation as presented in Eq. (21):

$$\underline{y} = \underline{\underline{A}}^{-1} \underline{c} \tag{21}$$

where A is a coefficient square matrix with dimension $(3N + 4S) \times (3N + 4S)$ composed by incidence matrix M, specification matrix E, identity matrix INN with dimension $N \times N$, identity matrix ISS with dimension $S \times S$, zero matrix ONN with dimension $N \times N$, zero matrix OSS with dimension $S \times S$, zero matrix ONS with dimension $N \times S$ and zero matrix OSN with dimension $S \times N$, as shown in Eq. (22):

$$\underline{\underline{A}} = \begin{bmatrix} \begin{bmatrix} \underline{\underline{M}} \\ \underline{\underline{0}}_{NS} \\ -\underline{\underline{F}}_L \\ \underline{\underline{0}}_{SS} \\ \underline{\underline{0}}_{SS} \end{bmatrix} & \begin{bmatrix} \underline{\underline{0}}_{NS} \\ \underline{\underline{M}} \\ -\underline{\underline{F}}_G \\ \underline{\underline{0}}_{SS} \\ \underline{\underline{0}}_{SS} \end{bmatrix} & \begin{bmatrix} \underline{\underline{0}}_{NS} \\ \underline{\underline{0}}_{NS} \\ \underline{\underline{I}}_{SS} \\ \underline{\underline{0}}_{SS} \\ -\underline{\underline{I}}_{SS} \end{bmatrix} & \begin{bmatrix} \underline{\underline{0}}_{NS} \\ \underline{\underline{0}}_{NS} \\ -\underline{\underline{I}}_{SS} \\ -\underline{\underline{I}}_{SS} \\ \underline{\underline{0}}_{SS} \end{bmatrix} & \begin{bmatrix} \underline{\underline{0}}_{NN} \\ \underline{\underline{0}}_{NN} \\ \underline{\underline{0}}_{SN} \\ \underline{\underline{M}}_2^T \\ \underline{\underline{M}}_1^T \end{bmatrix} & \begin{bmatrix} \underline{\underline{I}}_{NN} \\ \underline{\underline{0}}_{NN} \\ \underline{\underline{0}}_{SN} \\ \underline{\underline{0}}_{SN} \\ \underline{\underline{0}}_{SN} \end{bmatrix} & \begin{bmatrix} \underline{\underline{0}}_{NN} \\ \underline{\underline{I}}_{NN} \\ \underline{\underline{0}}_{SN} \\ \underline{\underline{0}}_{SN} \\ \underline{\underline{0}}_{SN} \end{bmatrix} \\ & & & \underline{\underline{E}} & & & \end{bmatrix} \quad (22)$$

and $\underline{c}$ is a vector with dimension $(3N + 4S) \times 1$ containing zero vectors $0N$ with dimension $N \times 1$, zero vectors $0S$ with dimension $S \times 1$, the non-homogeneous part of the system d with dimension $S \times 1$ and the specification vector e with dimension $(N + S) \times 1$ as shown in Eq. (23):

$$\underline{c} = \begin{bmatrix} \underline{0}_N^T & \underline{0}_N^T & \underline{d}^T & \underline{0}_S^T & \underline{0}_S^T & \underline{e}^T \end{bmatrix}^T \quad (23)$$

Numerical Methods

The methods necessary to solve the proposed model must attend the following tasks: solution of a system of linear algebraic equations, solution of a system of ordinary differential equations (ODE) and solution of a system of non-linear algebraic equations (NLE). For the solution of the ODE system it used a 3rd to 4th order Runge–Kutta type method. For the solution of the NLE system it implemented a proper routine based on Newton–Raphson method with numerical Jacobian, where the adopted convergence criterion was the norm of residue vector equal to 10^{-10}.

SIMULATION RESULTS

Two case studies are presented: (i) first case study analyzes an offshore petroleum well with gas lift, as a result it was possible to determine the behavior petroleum production as a function of the injected gas flow rate for different reservoir pressures and different wellbore diameters and (ii) second case study simulates a complex petroleum production system with multiple wells. For these simulations, the fluids properties are summarized in Table 1.

Table 1: Properties of fluids

Fluid	Liquid density (kg/m³)	Gas reference density (kg/m³)	Gas normal density (kg/m³)	Gas reference pressure (bar)	Gas polytrophic coefficient (–)	Viscosity (Pa s)	Interfacial tension (Pa m)
Oil	850.0	–	–	–	–	$1.2\ 10^{-2}$	0.071
Gas	–	80.0	1.2	100	1.25	$8.0\ 10^{-5}$	

Case 1—Offshore Single Well

In this section, the presented model is applied to evaluate the behavior of a subsea petroleum production pipeline network composed by a well connected to a production line with geometry described in Fig. 3. The topology of this network is composed by 9 nodes and 8 pipes according to the digraph presented in Fig. 4, where the fifth node corresponds to a well. The well is represented in this digraph as a square, and contains 6 nodes and 5 pipes, as shown in Fig. 5.

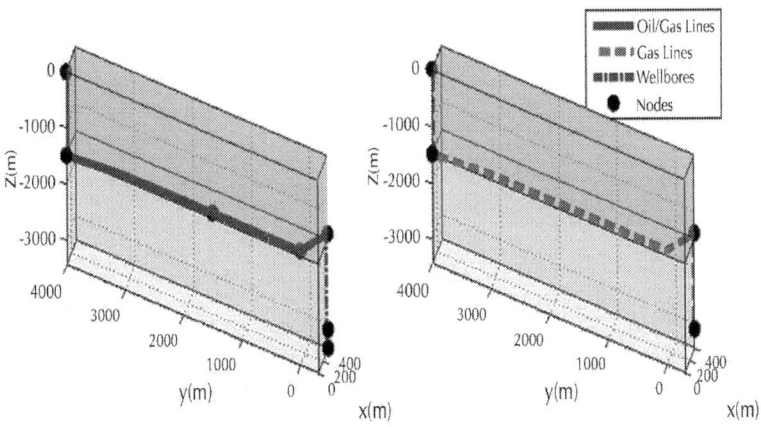

Figure 3: Case 1—simplified isometric projection.

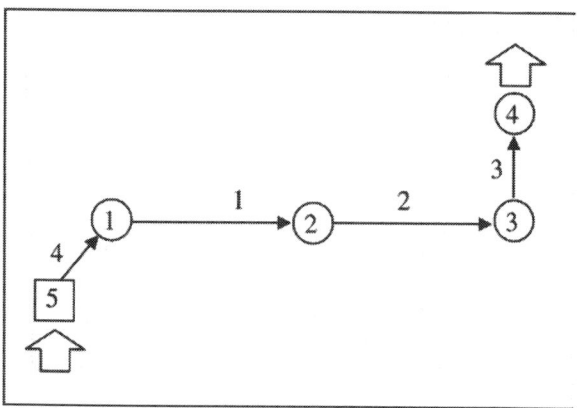

Figure 4: Case 1—network digraph.

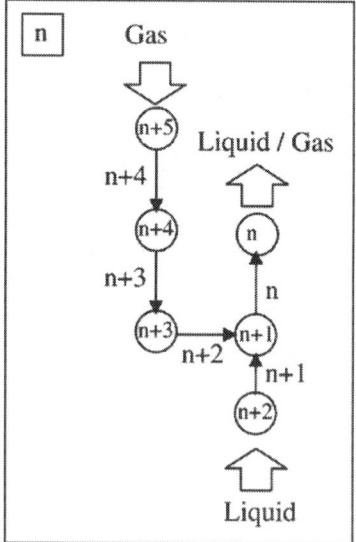

Figure 5: Well digraph.

The network lines are production line (1 and 2), riser (3), well to production line (4) and the following well lines (where n is equal to 5): wellbore (n and $n + 1$), gas lift annulus ($n + 2$) and gas supply lines ($n + 3$ and $n + 4$). The parameters used to describe these lines are shown in Table 2, Table 3 and Table 4.

Table 2: Case 1—network lines description

Line	Type	Diameter (in)	Length (m)	Inclination (°)	Roughness (μm)
1	Production line—part 1	6	1500	0°	100
2	Production line—part 2	6	4005	See Table 5	100
3	Riser	6	1500	90°	100
4	Well to production line	5	500	0°	100
5	Wellbore—part 2	5	1500	90°	100
6	Wellbore—part 1	5	300	90°	100
7	Gas lift annulus	a	1500	−90°	0
8	Gas line—part 1	2	4505	See Table 6	0
9	Gas line—part 2	2	1515	0°	0

[a] Hydraulic diameters: annulus with internal diameter 5 in and external diameter 10 in.

Table 3: Case 1—line 2 profile

Length (m)	0	1500	2505
Height (m)	−1400	−1400	−1500
x-coordinate (m)	0	0	0
y-coordinate (m)	1500	3000	4000

Table 4: Case 1—line 8 profile

Length (m)	0	1005	2505	**4005**	**4505**
Height (m)	−1495	−1395	−1395	−1395	−1400
x-coordinate (m)	0	0	0	0	400
y-coordinate (m)	4000	3000	1500	0	−300

This network has one source of liquid represented by node 7 at the deepest level of the wellbore and one source of gas at the platform represented by node 10. Node 4 represents the riser end (before the choke valve) and is specified with a fixed pressure. The other nodes interconnect the lines. The description of these nodes is presented in Table 5. The network nodal specifications are detailed in Table 6.

Table 5: Case 1—nodal description

Node	Description	X (m)	Y (m)	Height (m)
1	Production line inlet	0	0	−1400
2	Production line interconnection	0	1500	−1400
3	Production line—riser connection	0	4000	−1500
4	Riser outlet	0	4000	0
5	Wellhead	400	−300	−1400
6	Gas lift annulus and wellbore interconnection	400	−300	−2900
7	Downhole	400	−300	−3200
8	Annulus gas inlet	400	−300	−1400
9	Gas lines interconnection	0	4000	−1495
10	Gas inlet	0	4000	20

Table 6: Case 1—nodal specifications

Node	Type	Specification 1	Specification 2
1, 2, 3, 5, 6, 8 and 9	Connection	$w_L i = 0$ kg/s	$w_G i = 0$ kg/s
4	Fixed pressure	$PW_4 = 15$ bar	–
7	Oil inlet	w_{L7} equal to oil production	$w_{G7} = 0$ kg/s
10	Gas inlet	w_{G10} equal to gas injection	$w_{L10} = 0$ kg/s

The two-phase flow model for this network consists of 66 equations: 9 mass balance equations for liquid phase at lines, 9 mass balance equations for gas phase at lines, 9 momentum balance equations at pipes, 20 momentum balance equations at nodes and 19 specification equations.

Varying the injected gas mass flow rate and the produced oil mass flow rate, it is possible to generate surface graphs of all network variables. Important information is achieved when the contour plot of the downhole pressure is plotted, as shown in Fig. 6. Considering a fixed value of downhole pressure, it is possible to evaluate the required gas flow rate to elevate a certain quantity of petroleum. For instance, considering a downhole pressure of 220 bar, to produce 10.0 kg/s of petroleum it is necessary to inject 1.0 kg/s of gas or a higher value of gas flow rate, equal to 2.7 kg/s. Another value of importance for the well design is the maximum petroleum flow rate for a certain pressure, for instance, at downhole pressure of 220 bar, the maximum oil production is 10.7 kg/s with gas injection of 1.8 kg/s.

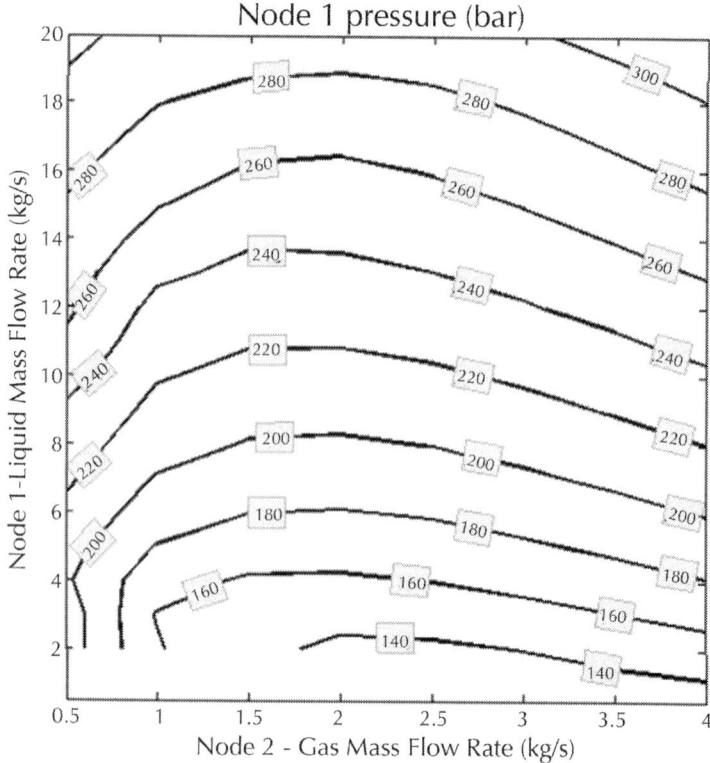

Figure 6: Case 1—downhole pressure contour plot.

The analysis of these results confirms that the model can predict behaviors commonly verified during petroleum elevation operations:

- Considering a fixed value of inlet pressure, the increase in the gas injection flow rate can increase the liquid flow rate
- There is a maximum liquid flow rate that is obtained with a critical gas flow rate. If the gas flow rate is increased beyond this critical value, a decrease in the liquid flow rate is observed.

Another important variable to be considered is the pressure at the compressor discharge or the pressure at the gas inlet (node 10). The results for this variable are presented in Fig. 7. It is possible to observe elevated values of pressures for high gas mass flow rates. For comparison, the simulation for gas lines (lines 8 and 9) with 3" diameter is presented in Fig. 8.

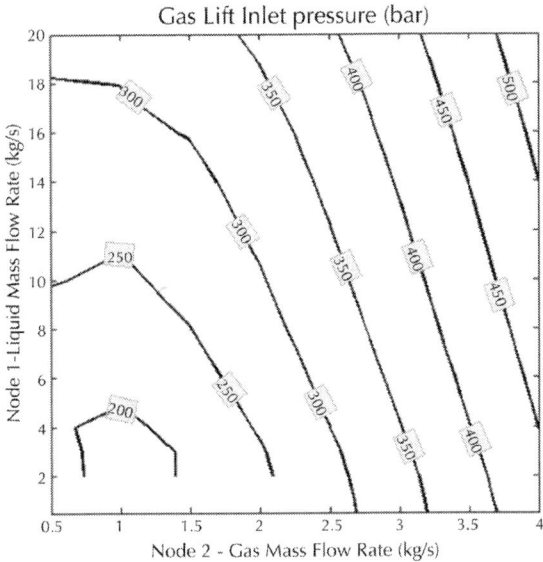

Figure 7: Case 1—gas lift inlet pressure contour plot (2″ gas line).

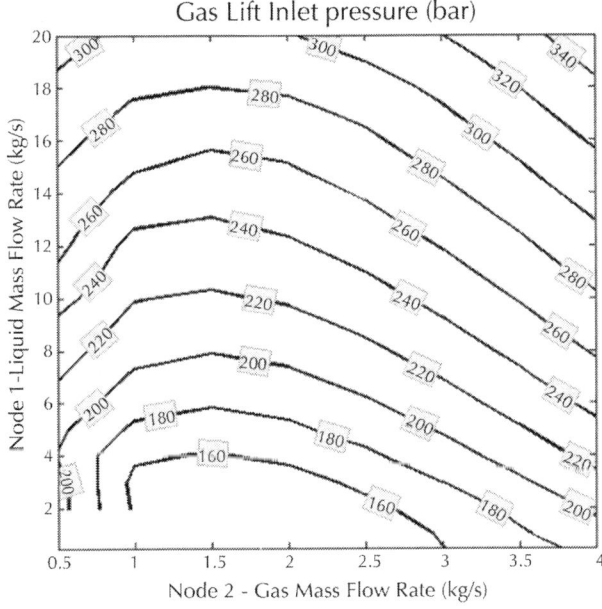

Figure 8: Case 1—gas lift inlet pressure contour plot (3″ gas line).

Case 2—Offshore Multiple Wells

In order to evaluate the application of the presented model to a complex subsea petroleum production pipeline network, this section presents a case study of a complex network with geometry described in Fig. 9. The topology of this network is composed of 28 nodes and 27 pipes according to the digraph in Fig. 10 (where the nodes 5, 11, 17 and 13 correspond to wells in which the structure is presented in Fig. 5).

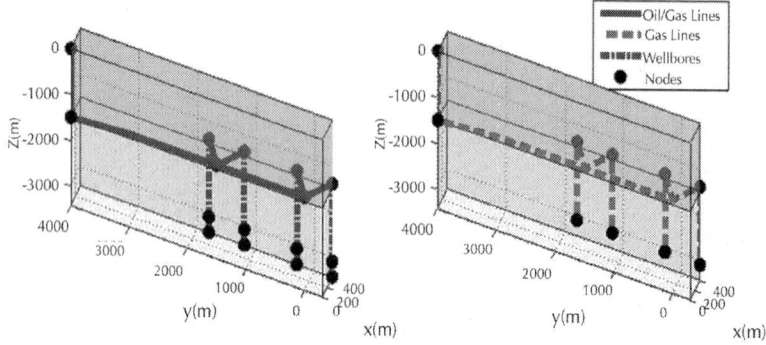

Figure 9: Case 2—simplified isometric projection.

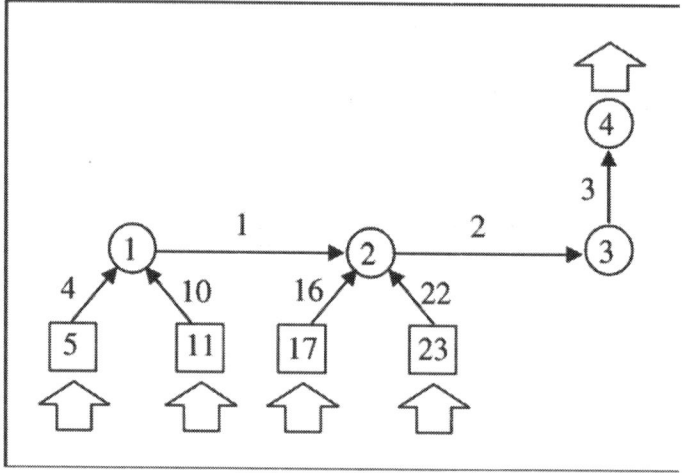

Figure 10: Case 2—network digraph.

The network lines are as follows: production line (1 and 2), riser (3), well to production line ($n - 1$), wellbore (n and $n + 1$), gas lift annulus ($n + 2$) and gas supply lines ($n + 3$ and $n + 4$), where n is equal to 5, 11, 17 and 23 (see Fig. 5). The parameters used to describe these lines are shown in the Table 7, Table 8, Table 9 and Table 10.

Table 7: Case 2—lines description

Line	Type	Diameter (in)	Length (m)	Inclination (°)	Roughness (μm)
1	Production line—part 1	6	1500	0°	100
2	Production line—part 2	6	4005	See Table 5	100
3	Riser	6	1500	90°	100
($n - 1$)	Well to production line	5	500	0°	100
(n)	Wellbore—part 2	5	1500	90°	100
($n + 1$)	Wellbore—part 1	5	300	90°	100
($n + 2$)	Gas lift annulus	a	1500	$-90°$	100
($n + 3$)	Gas line—part 1	2	4505	See Table 6	100
($n + 4$)	Gas line—part 2	2	1515	0°	100

[a] Hydraulic diameters: annulus with inner diameter 5 in and outer diameter 10 in.

Table 8: Case 2 — line 2 profile

Length (m)	0	1500	2505
Height (m)	—	—	—
	1400	1400	1500
x-coordinate (m)	0	0	0

| y-coordinate (m) | 1500 | 3000 | 4000 |

Table 9: Case 2—lines 8 and 14 profile

Length (m)	0	1005	2505	4005	4505
Height (m)	−1495	−1395	−1395	−1395	−1400
x-coordinate (m)	0	0	0	0	a
y-coordinate (m)	4000	3000	1500	0	a

[a] According to the correspondent wellhead coordinates.

Table 10: Case 2—lines 20 and 26 profile

Length (m)	0	1005	2505	4005
Height (m)	−1495	−1395	−1395	−1395
x-coordinate (m)	0	0	0	a
y-coordinate (m)	4000	3000	1500	a

[a] According to the correspondent wellhead coordinates.

This network has four sources of liquid represented by nodes 7, 13, 19 and 25 at the deepest level of the wellbore; and four sources of gas at the platform represented by nodes 10, 16, 22 and 28. Node 4 represents the riser end (before the choke valve) and is described with a fixed pressure. The other nodes interconnect lines. The description of these nodes and their specifications are presented in Table 11. The network nodal specifications are detailed in Table 12.

Table 11: Case 2—nodal description

Node	Description	X (m)	Y (m)	Height (m)
1	Production line inlet	0	0	−1400
2	Production line interconnection	0	1500	−1400
3	Production line—riser connection	0	4000	−1500
4	Riser outlet	0	4000	0

5	Wellhead 1	400	−300	−1400
11	Wellhead 2	400	300	−1400
17	Wellhead 3	400	1200	−1400
23	Wellhead 4	400	1800	−1400
(n + 1)	Gas lift annulus and wellbore interconnection	a	a	−2900
(n + 2)	Downhole	a	a	−3200
(n + 3)	Annulus gas inlet	a	a	−1400
(n + 4)	Gas lines interconnection	0	4000	−1495
(n + 5)	Gas inlet	0	4000	20

[a] Equal to the correspondent wellhead.

Table 12: Case 2—Nodal specifications

Node	Type	Specification 1	Specification 2
1, 2, 3, (n), (n + 1), (n + 3) and (n + 4)	Connection	$wLi = 0$ kg/s	$wGi = 0$ kg/s
4	Fixed pressure	$PW_4 = 15$ bar	–
(n + 2)	Oil inlet	w_Li: n-th well production	$wGi = 0$ kg/s
(n + 5)	Gas inlet	w_Gi: n-th well gas injection	$wLi = 0$ kg/s

The two-phase flow model for this network consists of 232 equations: 27 mass balance equations for liquid phase at lines, 27 mass balance equations for gas phase at lines, 27 momentum balance equations at pipes, 56 momentum balance equations at nodes and 95 specification equations.

Varying the injected gas mass flow rate from 0.5 to 4.0 kg/s and the produced oil mass flow rate from 1.0 to 20.0 kg/s, fixing the same injected gas and produced oil flow rates in each one of the wells, it is possible to generate surface graphs of all network variables. Important information is achieved when the contour plot of the downhole pressure is plotted, as shown in Fig. 11 for nodes 7 and 13 and in Fig. 12 for nodes 19 and 25.

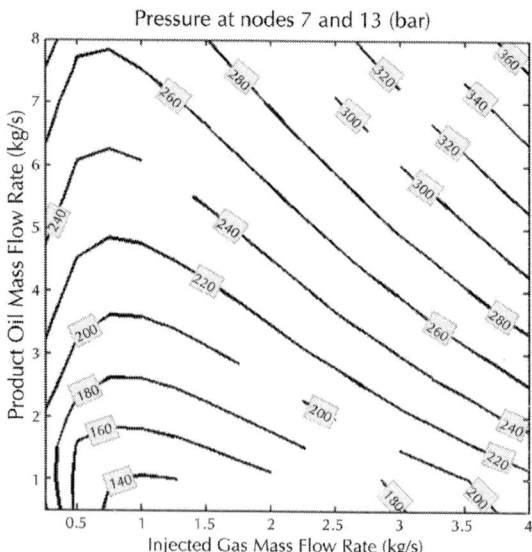

Figure 11: Case 2—downhole pressure contour plot (nodes 7 and 13).

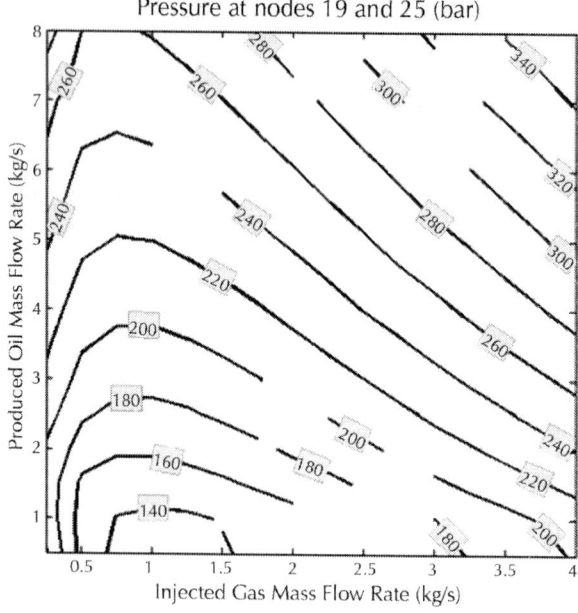

Figure 12: Case 2—downhole pressure contour plot (nodes 19 and 25).

GAS LIFT OPTIMIZATION

Problem Formulation

As described previously, this paper proposes an optimization framework to determine the optimal operational conditions and/or the optimal design of a gas lift system for offshore petroleum production systems. This optimization problem can be described by Eq. (24):

$$\text{Maximize}_{\underline{u} \in \mathbb{R}^m} \Omega_j$$

$$\text{subjected to}: \begin{cases} \underline{LB} < \underline{u} < \underline{UB} \\ (\underline{P}_{well} - \underline{P}_{downhole}) - \text{diag}^{-1}(\underline{PI})\underline{Q}_L = \underline{0} \end{cases} \quad (24)$$

where Ωj is the j-type objective function, u is a vector of m decision variables, uop is a solution vector, LB is a vector of lower bounds, UB is a vector of upper bounds, P_{well} is the vector of reservoir pressures in bar, $P_{downhole}$ is the vector of pressures at the wellbore bottom in bar, PI is the vector of productivity index in (kg/s)/bar and QL is the vector of liquid production in kg/s.

The vector of decision variables u can be composed by three types of variables:

- Injected gas flow rate denoted by the vector Q_G (where $Q_G i$ corresponds to the mass flow rate of gas in kg/s injected in the i-th well)
- Produced liquid flow rate denoted by the vector QL (where $Q_L i$ corresponds to the mass flow rate of liquid in kg/s produced by the i-th well)
- Diameters of gas pipelines denoted by the vector D_G (where $D_G i$ corresponds to the diameter in inches of the gas lines that transport the gas from the platform to the well annuli).

The objective function Ωj can be formed by three terms in million USD/yr corresponding to annualized revenue (R, in this case petroleum production), annualized utilities cost (A_{GC}, in this case only fuel gas is considered) and annualized total capital cost (A_{TCC}).

Thus, three different types of objective functions are obtained:

(1) Maximum production (objective function Ω_1): determination of the best operational conditions and the injected gas that guarantee the maximum oil production for a fixed productivity index.

$$\Omega_1 = R \qquad (25)$$

(2) Maximum profit (objective function Ω_2): determination of the injected gas that guarantee the maximum profit considering the fuel gas cost and the oil price for a fixed productivity index.

$$\Omega_2 = R - A_{GC} \qquad (26)$$

(3) Optimal design (objective function Ω_3): determination of the injected gas and the gas pipe diameters that guarantee the optimal design of a continuous gas lift system, considering the capital costs with compressor and gas pipelines, installation and maintenance costs, fuel gas cost and the petroleum price.

$$\Omega_3 = R - A_{GC} - A_{TCC} \qquad (27)$$

The inside battery limits cost, *ISBL*, determined by compressor, gas turbine and gas pipelines installed cost, is calculated according to Eq. (28)

$$ISBL = C_C + C_T + C_P \qquad (28)$$

where C_C is the compressor installed cost, C_T is the turbine installed cost and C_P is the gas pipelines installed cost (each one in millions USD).

The annualized total capital cost A_{TCC} can be calculated via Eq. (29):

$$A_{TCC} = CCF \; OSBL \tag{29}$$

where *CCF* can be estimated as 0.333 yr^{-1} (Douglas, 1988) considering an interest rate of 15% and 12 years of amortization.

The outside battery limits cost, *OSBL,* can be estimated for chemical process plants as described in Douglas (1988) as approximately three times the *ISBL*.

Compressor Capital Cost Estimation

Considering a centrifugal compressor, the cost is presented by Couper et al. (2005) as described in Eq. (30):

$$C_C = 14.3 \cdot 10^{-3} w_C^{0.62} \tag{30}$$

where w_C is the power of the machine in kW.

The work of compression is calculated considering ideal gas in an *n*-stage reciprocating machine with complete intercooling using Eq. (58) described in Perry and Green (1997)

$$w_C = \frac{n\gamma RT_{IN}}{\eta(\gamma-1)MW} \left(\left(\frac{\max(P_G)}{P_{IN}} \right)^{\frac{\gamma-1}{n\gamma}} - 1 \right) (\underline{Q_G}^T \underline{1}) \tag{31}$$

where T_{IN} and P_{IN} are the gas temperature and pressure, respectively, at the compressor suction in bar, P_G is a vector of pressures at the gas pipeline inlets in bar, Q_G is the vector of injected gas in kg/s, *MW* is the molecular weight of the gas in kg/kmol, *n* is the number of compression stages, is the compression efficiency and *R* is the universal gas constant (8.314 kJ/kmol/K).

Gas Turbine Capital Cost Estimation

Considering a centrifugal compressor with a gas turbine motor drive, the cost can be estimated according to Couper et al. (2005) as described in Eq. (32):

$$C_T = 7.19 \cdot 10^{-4} w_C^{0.81}$$

(32)

where w_C is the power of the machine in kW.

Capital Cost Estimation with Gas Pipes

The capital cost related to the acquisition of gas pipelines to transport the injected gas from the compressor discharge to the well can be described by Eq. (33) according to the work of McCoy (2008) for USA Southwest cost parameters and not considering right-of-way costs.

$$C_P = \sum_i \left(\begin{array}{l} 1.29 \cdot 10^{-3} D_{G,i}^{1.59} \left(\dfrac{L_{G,i}}{1000} \right)^{0.901} + \\ + 18.7 \cdot 10^{-3} D_{G,i}^{0.940} \left(\dfrac{L_{G,i}}{1000} \right)^{0.820} + \\ + 24.5 \cdot 10^{-3} D_{G,i}^{0.791} \left(\dfrac{L_{G,i}}{1000} \right)^{0.783} \end{array} \right)$$

(33)

Where CP is the installed cost in millions USD, L_G,i is the gas pipe length from the platform to the i-th well in m and D_G,i is the diameter of the gas pipelines of i-th well in inches.

Revenue and Operating Cost Estimation

The variable cost of fuel gas, considering an average efficiency for the gas turbine and the compressor, is determined by:

$$A_{GC} = 8.45 \cdot 10^{-3} w_C C_{GC}$$

(34)

where A_{GC} is the fuel gas cost in million USD/yr, w_C is the compressor power in kW and c_{GC} is the gas cost in USD/m³.

The revenue generated by the petroleum production is determined by Eq. (35):

$$R = 198.34 \frac{\left(Q_l^T 1\right)}{\rho_{LO}} c_L \qquad (35)$$

where R is the revenue value in million USD/yr, ρ_{LO} is the oil density at standard conditions in kg/m³, and c_L is the petroleum price in USD/barrel.

Connectivity between Simulation and Optimization Algorithms

Before describing the connectivity between the algorithms, it is important to present the set of: (i) specifications and parameters for the pipe network model; (ii) decision variables and parameters for the optimization algorithm; and (iii) all data transferred between each part.

The stationary two-phase flow model for pipe networks, described in Section 3, requires parameters related to pipe geometry (length, diameter and roughness), fluid properties (density, viscosity and compressibility) and (iii) digraph connectivity, and specifications that, for the proposed approach, are nodal variables (pressures, external liquid mass flow rates and/or external gas mass flow rates), where the number of specifications depends on the number of pipes and nodes.

For the scope of this work, a special type of pipe network for petroleum production is considered – convergent tree-type networks – where each node can be classified as follows:

- Oil source: specified with a non-zero liquid mass flow rate and a null gas mass flow rate (corresponds to the well downhole at the reservoir);
- Gas source: specified with a null liquid mass flow rate and a non-zero gas mass flow rate (corresponds to the injected gas entrance at the platform);
- Connections: specified with null liquid and gas mass flow rates; and
- Fixed pressure: specified with a fixed value of nodal pressure (corresponds to a certain point before the choke valve at the platform).

If the specifications are adequate in terms of physical meaning, the solution of the two-phase flow network model (Eqs. (6) and (14)) generates some important variables that are used to compose the optimization problem, like: pressure at oil sources (downhole pressures, $P_{downhole}$); pressure at gas sources (injected gas pressure at the platform, P_G); and pressure at connections.

The commitment of the optimization algorithm to the simulation algorithm is to specify mass flow rates of the Oil Source nodes and the Gas Source nodes, Q_L and Q_G, respectively, based on the decision variables vector u.

The commitment of the simulation algorithm to the optimization algorithm is to calculate pressures at the Oil Source nodes and the Gas Source nodes, that corresponds to the $P_{downhole}$ (pressure at the downhole) and the P_G (injected gas pressure), respectively, required by Eqs. (24) and (31).

Numerical Methods

The optimization method employed is based on a sequential quadratic programming (SQP) algorithm. Constraints on the decision variables are applied to avoid unfeasible specifications: (i) diameter lower bound equal to 1 inch, (ii) diameter upper bound equal to 4 in, (iii) minimum gas injection equal to 0 kg/s and (iv) minimum oil production equal to 0 kg/s.

OPTIMIZATION RESULTS

In 6.1 and 6.2, the gas lift optimization algorithm will be applied for two different cases (offshore single well case and offshore multiple wells, with geometry and other specifications described) that have geometries and simulation results properly presented in 5.1 and 5.2, respectively.

Case 1—Offshore Single Well

Maximum Production

The purpose of this subsection is to test and exemplify the application of optimization algorithms for the determination of the optimal injected gas mass flow rate that guarantee the maximum oil production for an offshore single well system (with optimizations repeated for different downhole pressure specifications). The objective function Ω_1 is maximized considering different downhole pressures varying from 180 to 260 bar (*P*well in Eq. 24) and an initial estimate of 10 kg/s of produced oil (*QL* in Eq. 35) and 2 kg/s of injected gas (*QG* in Eq. 31). For these scenarios the productivity index (*PI* in Eq. 24) is equal to infinite, that corresponds to downhole pressure equal to reservoir pressure. The solutions obtained are summarized in Fig. 13 and Fig. 14. It is verified that the optimal flow rate of gas injection varies from 1.55 to 1.85 kg/s.

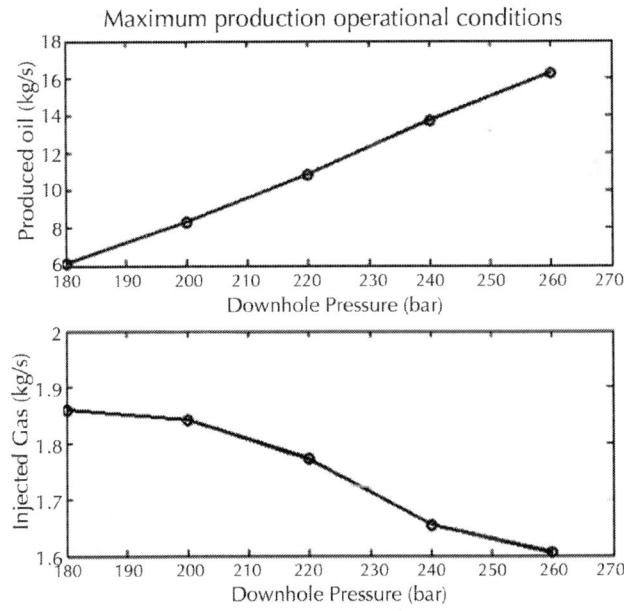

Figure 13: Case 1—maximum production for different well pressures.

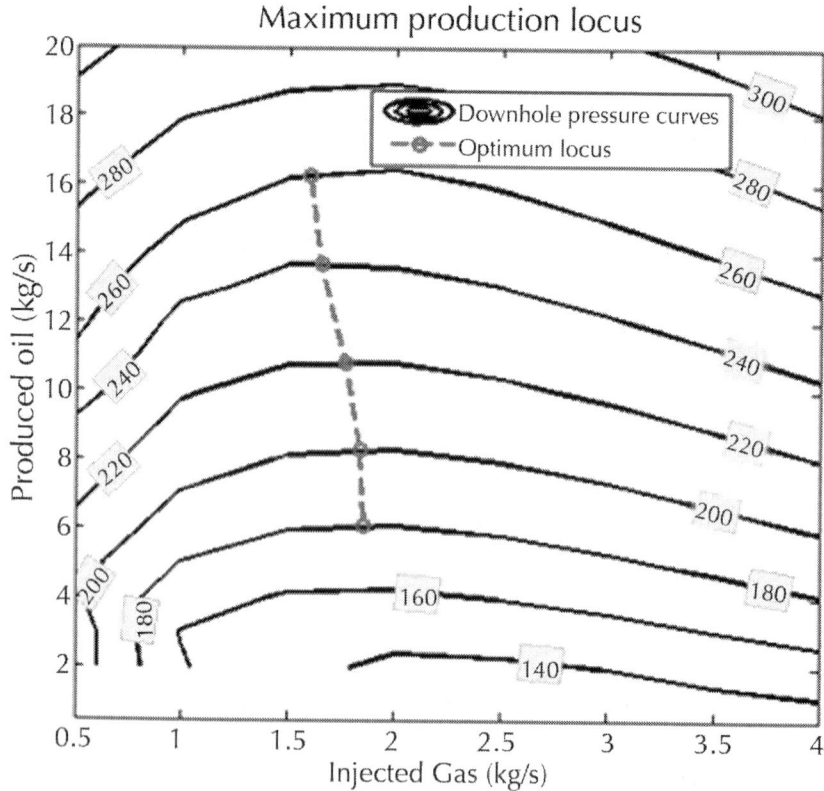

Figure 14: Case 1—maximum locus for different well pressures.

Maximum Production Considering Different Productivity Indexes

The objective function Ω_1 is maximized considering reservoir pressure (P_{well} in Eq. 24) equal 220 bar, different productivity index (PI in Eq. 24) varying from 0.5 to 10 (kg/s)/bar and an initial estimate of 10 kg/s of produced oil (Q_L in Eq. 35) and 2 kg/s of injected gas (QG in Eq. 31). The solutions obtained are summarized in Fig. 15.

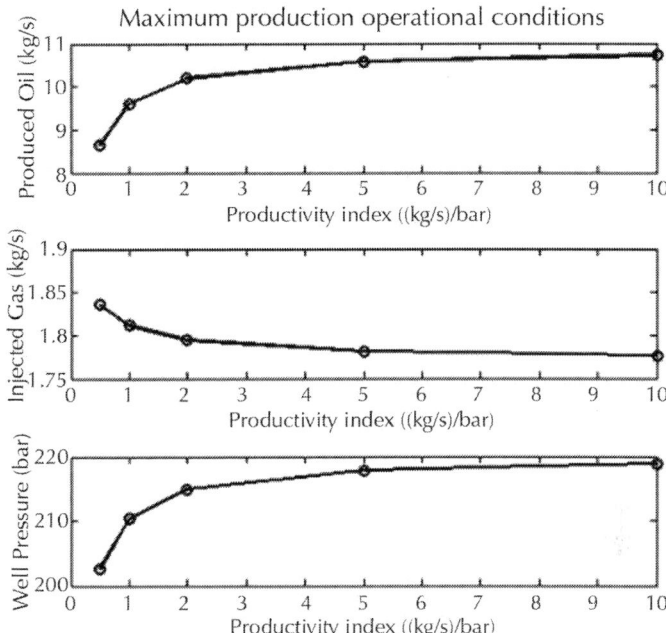

Figure 15: Case 1—maximum production for different productivity index.

A common behavior for petroleum wells corresponds to the reduction of productivity index across the years of production. According to Fig. 15, the flow rate of injected gas shall be adjusted to guarantee the maximum production during over the well life.

Maximum Profit

The profit objective function Ω_2 is maximized for the same well conditions considering different oil prices. With the results presented in Fig. 16 and detailed in Table 13, it is possible to determine the operational conditions according to the petroleum price in comparison to the fuel gas price. Analyzing the results of Table 13, it verified a reduction in the oil production in scenarios of lower petroleum prices. In the proposed case, the impact of the cost ratio of fuel gas and oil price is low, due to the small cost of fuel gas in comparison to the amount of gain with the produced oil.

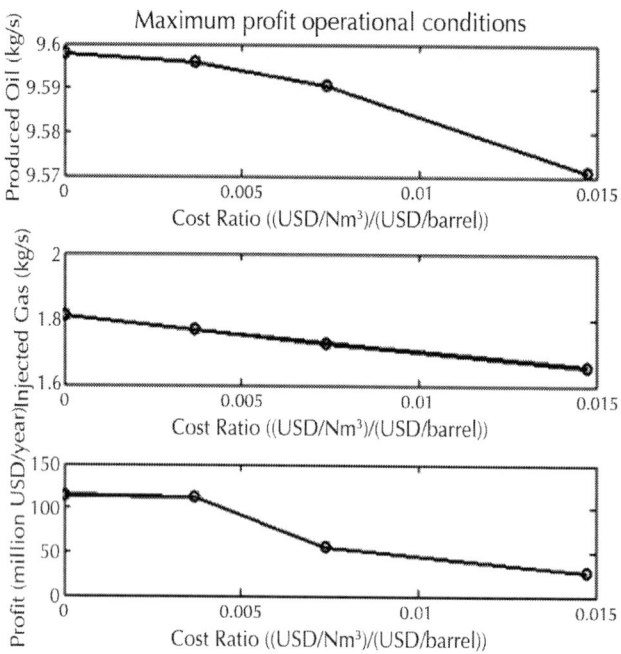

Figure 16: Case 1—maximum profit for different price scenarios.

Table 13: Case 1—maximum profit results

Scenario	1	2	3	Max
Cost ratio 10^3((USD/Nm^3)/(USD/barrel))	14.74	7.37	3.69	0
Fuel gas cost (USD/Nm^3)	0.1886	0.1886	0.1886	0
Oil price (USD/barrel)	12.79	25.58	51.15	51.15
Reservoir pressure (bar)	220	220	220	220
Productivity index ((kg/s)/bar)	1.0	1.0	1.0	1.0
Produced oil (kg/s)	9.571	9.591	9.596	9.598
Injected gas (kg/s)	1.658	1.730	1.770	1.812
Downhole pressure (bar)	210.43	210.41	210.40	210.40
Compressor power (kW)	1 002.7	1 051.8	1 078.9	1 108.1
Discharge pressure (PG)	269.18	274.03	276.74	279.68

| Profit (10⁶ USD/yr) | 26.96 | 55.56 | 112.82 | 114.56 |

Case 2—Offshore Multiple Wells

Maximum Production Considering Productivity Index

The objective function Ω_1 is maximized considering an initial estimate of 3 kg/s of produced oil (Q_L in Eq. 35) and 1 kg/s of injected gas (Q_G in Eq. 31) for each well. Two scenarios are analyzed varying the reservoir pressure. The solutions obtained are described in Table 14 and Table 15.

Table 14: Case 2—maximum production results for a reservoir with 220 bar

Well	1	2	3	4
Node	7	13	19	25
Productivity index ((kg/s)/bar)	1.0	2.5	1.0	0.5
Downhole pressure (bar)	215.80	218.03	214.85	211.61
Produced oil (kg/s)	4.204	4.933	5.152	4.197
Injected gas (kg/s)	0.743	0.822	0.897	0.795

Table 15: Case 2—Maximum production results for a reservoir with 240 bar

Well	1	2	3	4
Node	7	13	19	25
Productivity index ((kg/s)/bar)	1.0	2.5	1.0	0.5
Downhole pressure (bar)	234.58	237.43	233.44	229.45
Produced oil (kg/s)	5.417	6.423	6.561	5.276
Injected gas (kg/s)	0.681	0.744	0.820	0.734

Analyzing the results of the optimization of multiple wells systems for maximum production, the following considerations can be stated:

- For wells with equal productivity index and similar geometry (wells 1 and 3), more gas is injected at wells that are closer to the platform and consequently these wells are more productive.
- For wells at similar distances and geometry (wells 3 and 4), more gas is injected for wells with higher productive index.
- Comparing scenarios 1 and 2, for higher reservoir pressure a larger amounts of oil are produced and less injected gas is required.

Optimal Design

The application of the objective function Ω_3 to the offshore multiple wells case with the design parameters presented in Table 16, has the objective of determining the optimal design considering the annualized costs of compressor, turbine driver and gas pipelines (that transport gas from the compressor discharge to the well annulus) and fuel gas.

Table 16: Case 2—design parameters

Parameters	Value	Unit
Number of compressor stages (n)	3	–
Overall compressor efficiency (η)	0.8	–
Fuel gas polytrophic coefficient ( )	1.25	–
Injected gas temperature at suction (T_{IN})	300	K
Injected gas pressure at suction (PIN)	5	bar
Injected gas molecular weight (MW)	25	kg/kmol
Fuel gas cost (c_G)	0.189	USD/m³
Oil price (c_L)	51.15	USD/barrel
Pipe length (L_{G1})—lines 8 and 9	6020	m
Pipe length (L_{G2})—lines 14 and 15	6020	m
Pipe length (L_{G3})—lines 20 and 21	4520	m
Pipe length (L_{G4})—lines 26 and 27	4520	m
Diameter lower bound (LB)	1	in
Diameter upper bound (UB)	4	in
Productivity index—well 1	1.0	(kg/s)/bar
Productivity index—well 2	2.5	(kg/s)/bar
Productivity index—well 3	1.0	(kg/s)/bar
Productivity index—well 4	0.5	(kg/s)/bar

Reservoir pressure—scenario 1	220	bar
Reservoir pressure—scenario 2	240	bar

The decision variables are the produced oil flow rates in each well (w_L of nodes 7, 13, 19 and 25), the injected gas flow rate in each well (w_G of nodes 10, 16, 22 and 28), the diameter of gas pipelines of well 1 (lines 8 and 9), well 2 (lines 14 and 15), well 3 (lines 20 and 21) and well 4 (lines 26 and 27).

The initial estimates of decision variables are obtained from solution of the maximum production optimization presented in Table 14 and gas pipe diameters equal to 2 in.

The results obtained for the optimal design are presented in Table 17, Table 18 and Table 19.

Table 17: Case 2—decision variables for optimal design

Decision variables	Scenario 1	Scenario 2	Unit
Injected gas flow rate (Q_{G1})—well 1	0.706	0.647	kg/s
Injected gas flow rate (Q_{G2})—well 2	0.785	0.709	kg/s
Injected gas flow rate (Q_{G3})—well 3	0.854	0.780	kg/s
Injected gas flow rate (Q_{G4})—well 4	0.752	0.697	kg/s
Oil flow rate (Q_{L1})—well 1	4.197	5.411	kg/s
Oil flow rate (Q_{L2})—well 2	4.952	6.444	kg/s
Oil flow rate (Q_{L3})—well 3	5.152	6.559	kg/s
Oil flow rate (Q_{L4})—well 4	4.170	5.246	kg/s
Pipe diameter (D_{G1})—lines 8 and 9	1.77	1.66	in
Pipe diameter (D_{G2})—lines 14 and 15	1.89	1.76	in
Pipe diameter (D_{G2})—lines 20 and 21	1.75	1.63	in
Pipe diameter (D_{G4})—lines 26 and 27	1.62	1.52	in

Table 18: Case 2—pipelines, compressor and turbine optimal design results

Results	Scenario 1	Scenario 2	Unit
Compressor power (w_c)	1 797.5	1 691.1	kW
Compressor discharge pressure	235.44	258.02	bar
Compressor capital cost (C_c)	1.79	1.72	10^6 USD
Turbine capital cost (C_T)	3.95	3.76	10^6 USD
Gas pipelines capital cost (C_p)	1.05	0.98	10^6 USD
Fuel gas flow rate	15.19	14.29	10^6 m³/yr

Table 19: Case 2—optimal design results

Results	Scenario 1	Scenario 2	Unit
Total injected gas flow rate	81.39	74.45	10^6 m³/yr
Total oil flow rate	4.310	5.521	10^6 barrel/yr
Fuel gas cost (A_{GC})	2.86	2.69	10^6 USD/yr
Annualized capital cost	6.79	6.46	10^6 USD/yr
ISBL	6.79	6.46	10^6 USD
OSBL	20.36	19.38	10^6 USD
Revenue (R)	220.47	282.41	10^6 USD/yr
Profit	210.82	273.26	10^6 USD/yr

Economic parameters used in Eqs. (21), (23) and (24) are susceptible to deviations according to the market and the complexity of the enterprise. Furthermore, the major petroleum companies treat real economic data for offshore equipment, pipelines and labor costs as business secret. These difficulties forbade the presentation of the proposed methodology for a real case.

It is important to reinforce that a more complete analysis and, consequently, a more detailed objective function shall be applied to generate results that are more precise during the life cycle of an engineering design. The presented methodology and its objective function have an adequate precision to answer important questions of the conceptual design of a new enterprise.

CONCLUSIONS

This paper presents a NLP optimization coupled to a two-phase flow network model constituting a framework capable of answering three very common problems of petroleum companies: the maximum production estimation, the gas lift optimal operation conditions and the optimal design of continuous gas lift systems. The methodology permits the analysis of complex production system including multiple wells sharing the same production line.

This paper also presents a network formulation for a classical two-phase phase flow correlation – Beggs and Brill correlation – which is able to determine pressure gradient and hold-up in pipes with any inclination and under any flow pattern. The major motivation is a lack in the literature of works that describes the application of two-phase flow models to pipe networks. Another advantage is the mathematical simplicity that guarantees accurate results with short CPU time if compared with commercial software, corroborating its application in optimization purposes. For complex pipe networks composed by cycles and/or divergent networks, a simple modification to the proposed model is necessary to include the nodal hold-up and impose equal hold-ups in split nodes (where two or more pipes leave the same node).

Two different case studies exemplify the application of the optimization algorithm. The first case study, offshore single well case, determines the operational conditions that guarantee different objectives: maximum oil production for different well pressures, the optimal profit for different oil prices and the optimal design considering capital costs with compression unit and gas pipes. The second case study extends the analysis to a complex subsea system.

MATHEMATICAL NOTATION

Vectors are represented by underlined letters (u, U) and are defined as column vector. Matrices are represented by double underlined letters ($\underline{v}, \underline{\underline{V}}$). The transpose of a matrix $\underline{\underline{V}}$ is represented by the superscript T as $\underline{\underline{V}}^T$. The dimension of vectors and matrices is described using bold letters, for instance, the dimension of $\underline{\underline{M}}$ is N × S if $\underline{\underline{M}}$ has N row and S columns.

The × denotes element-by-element multiplication (the *Hadamard* or *Schur* product). The operator *diag*(*u*) generates a diagonal matrix $\underline{\underline{V}}$ with diagonal elements equal to the elements *u*. The *diag*$^{-1}$(*u*) operator generates a diagonal matrix $\underline{\underline{V}}$ with diagonal elements equal to the inverse of the elements of *u*. The operator *max*(*u*) returns the maximum element of *u*. The operator *max*(*a,b*) returns *a* if *a* > *b*, otherwise returns *b*.

ACKNOWLEDGMENT

This work was supported by Petróleo Brasileiro S.A. (PETROBRAS).

REFERENCES

1. Abel Waly, A.A., El-Massry, Y., Darweesh, T.A., Sallaly, M.El., 1996. Network model for an integrated production system applied to the Zeit Bay field, Egypt. J. Petrol. Sci. Eng. 15, 57–68.
2. Alarcon, G., Torres, C., Gomez, L., 2002. Global optimization of gas allocation to a group of wells in artificial lift using nonlinear constrained programming. J. Energy Resour. Technol. 124, 262–268.
3. Ansari, A.M., Sylvester, N.D., Sarcia, C., Shoham, O., Brill, J.P., 1994. A comprehensive mechanistic model for upward two-phase flow in wellbores. Proc. of 6th Annual SPE Meeting, New Orleans, Louisiana, USA, pp. 143–152.
4. Aziz, K., Govier, G.W., Fogarasi, M., 1972. Pressure drop in wells producing oil and gas. J. Can. Petrol. Technol. 11, 38–48.
5. Beggs, H.D., Brill, J.P., 1973. A study of two-phase flow in inclined pipes. J. Petrol. Technol. 607–617.
6. Camponogara, E., Nakashima, P., 2006. Optimizing gas lift optimization problem using dynamic programming. Eur. J. Oper.

Res. 174, 1220–1246.

7. Costa, A.L.H., Medeiros, J.L., Pessoa, F.L.P., 1998. Steady-state modeling and simulation of pipeline networks for compressible fluids. Braz. J. Chem. Eng., São Paulo 15 (4), 344–357.

8. Couper, J.R., Penney, W.R., Fair, J.R., Walas, S.M., 2005. Chemical Process Equipment, Selection and Design, Second Edition. Gulf Professional Publishing. 726 pp.

9. Douglas, J., 1988. Conceptual Design of Chemical Processes, 1st edition. McGraw-Hill Science/Engineering/Math.

10. Dutta-Roy, K., Kattapuram, J., 1997. A new approach to gas-lift allocation optimization. SPE 38333 presented at the SPE Western Regional Meeting.

11. Eaton, B.A., Andrews, D.E., Knowles, C.R., Silderberg, I.H., Brown, K.E., 1967. The prediction of flow patterns, liquid holdup and pressure losses ocurring during continuous two-phase flow in horizontal pipelines. J. Petrol. Technol. 12, 815–828.

12. Fang, W.Y., Lo, K.K., 1996. A generalized well-management scheme for reservoir\ simulation. SPE Reserv. Eng. 11, 116–120.

13. Flanigan, O., 1958. Effect of uphill flow on pressure drop in design of two-phase gathering systems. Oil Gas J. 56, 132–141.

14. Floquet, P., Joulia, X., Vacher, A., Gainville, M., Pons, M., 2009. Numerical and computational strategy for pressure-driven steady-state simulation of oilfield production. Comput. Chem. Eng. 33, 660–669.

15. Hagedorn, A.R., Brown, K.E., 1965. Experimental study of pressure gradients occurring during continuous two-phase flow in small-diameter vertical conduits. J. Petrol. Technol. 17 (4), 475–484.

16. Hasan, A.R., Kabir, C.S., 1992. Two-phase flow in vertical and inclined annuli. Int. J. Multiphase Flow 18 (2), 279–293.

17. Hughmark, G.A., 1962. Holdup in gas–liquid flow. Chem. Eng. Prog. 53, 62–65. Kjärstad, J., Johnsson, J., 2009. Resources and future supply of oil. Energy Policy 37 (2), 441–464.

18. Kosmidis, V., Perkins, J., Pistikopoulos, E., 2005. A mixed integer optimization formulation for the well scheduling problem on petroleum fields. Comput. Chem. Eng. 29 (7), 1523–1541.

19. Lockhart, R.W., Martinelli, R.C., 1949. Proposed correlation

of data for isothermal twophase two-component flow in pipes. Chem. Eng. Prog. 45, 39–48.
20. Mah, R.S.H., 1990. Chemical process structures and information flows, Series in Chemical Engineering, 1st ed. Butterworths.
21. McCoy, A.T., 2008. The Economics of CO2 Transport by Pipeline and Storage in Saline Aquifers and Oil Reservoirs. Ph.D. Thesis, Carnegie Mellon University, Pittsburgh, USA.
22. Mukherjee, H., Brill, J.P., 1985. pressure drop correlations for inclined two-phase flow. J. Energy Resour. Techno. 1, 1003–1008.
23. Oliemans, R.V.A., 1976. Two-phase in gas-transmission pipelines. 76-Pet-25, Joint Petroleum Mechanical Engineering & Pressure Vessels and Piping Conference, Mexico City, Mexico. September.
24. Oliemans, R.V.A., 1987. Modelling of gas condensate flow in horizontal and inclined pipes. Paper presented at the 1987 ASME Pipeline Engineering Symposium-ETCE, Dallas, TX.
25. Orkiszewski, J., 1967. Predicting two-phase pressure drops in vertical pipe. J. Petrol. Technol. 19, 829–838.
26. Perry, R.H., Green, D.W., 1997. In: Perry, R.H., Green, D.W. (Eds.), Perry's Chemical Engineers' Handbook, 7th ed. McGraw-Hill, New York, USA.
27. Ray, T., Sarker, R., 2007. Genetic algorithm for solving a gas lift optimization problem. J. Petrol. Sci. Eng. 59, 84–96.
28. Taitel, Y., Barnea, D., 1990. A consistent approach for calculating pressure drop in inclined slug flow. Chem. Eng. Sci. 45, 1199–1206.
29. Taitel, Y., Dukler, A.E., 1976. A model for predicting flow regime transitions in horizontal and near-horizontal gas–liquid flow. AIChE J. 22, 47–55.

Citations

CHAPTER 1

Chuanliang Yan, Jingen Deng, Lianbo Hu, and Baohua Yu, "Fracturing Pressure of Shallow Sediment in Deep Water Drilling," Mathematical Problems in Engineering, vol. 2013, Article ID 492087, 8 pages, 2013. doi:10.1155/2013/492087

CHAPTER 2

Daidai Wu, Nengyou Wu, Ying Ye, et al., "Early Diagenesis Records and Pore Water Composition of Methane-Seep Sediments from the Southeast Hainan Basin, South China Sea," Journal of Geological Research, vol. 2011, Article ID 592703, 10 pages, 2011. doi:10.1155/2011/592703.

CHAPTER 3

Lamendella R, Strutt S, Borglin S, Chakraborty R, Tas N, Mason OU, Hultman J, Prestat E, Hazen TC and Jansson JK (2014) Assessment of the Deepwater Horizon oil spill impact on Gulf coast microbial communities. Front. Microbiol. 5:130. doi: 10.3389/fmicb.2014.00130.

CHAPTER 4

Mohamed O. Abouelresh, "Multiscale Erosion Surfaces of the Organic-Rich Barnett Shale, Fort Worth Basin, USA," Journal of Geological Research, vol. 2013, Article ID 759395, 16 pages, 2013. doi:10.1155/2013/759395.

CHAPTER 5

Daniel Pipa, Sérgio Morikawa, Gustavo Pires, Claudio Camerini and oãoMárcio Santos, Flexible Riser Monitoring using Hybrid Magnetic/Optical Strain Gage Techniques through RLS Adaptive Filtering, doi:10.1155/2010/176203.

CHAPTER 6

X. Dong, Z. Zhou and H. Li, "Improve the Government Strategic Petroleum Reserves," Advances in Chemical Engineering and Science, Vol. 3 No. 4A, 2013, pp. 1-5. doi: 10.4236/aces.2013.34A1001.

CHAPTER 7

M. F. Fingas, "Studies on the Evaporation Regulation Mechanisms of Crude Oil and Petroleum Products,"AdvancesinChemicalEng ineeringandScience,Vol.2No.2,2012,pp.246-256.doi:10.4236/aces.2012.22029.

CHAPTER 8

J. Saraiva de Souza, T. MotaVieira, E. Santos Barbosa, A. de Lima Cunha, S. Rodrigues de Farias Neto and A. Gilson Barbosa de Lima, "Numerical Study of Oil/Water Separation by Ceramic Membranes in the Presence of Turbulent Flow," Advances in Chemical Engineering and Science, Vol. 2 No. 2, 2012, pp. 257-265. doi:10.4236/aces.2012.22030.

CHAPTER 9

J.N.M. de Souza, J.L. de Medeiros, A.L.H. Costa, G.C. Nunes, Modeling, simulation and optimization of continuous gas lift systems for deepwater offshore petroleum production, Journal of Petroleum Science and Engineering, Volume 72, Issues 3–4, June 2010, Pages 277-289, ISSN 0920-4105, http://dx.doi.org/10.1016/j.petrol.2010.03.028

Index

A
Adams SW#7 (ASW) 85

B
Bottom simulating reflector (BSR) 25
Boundary-layer 172, 173, 175, 185, 186, 187, 188, 189, 190

C
Continuous Gas Lift (CGL) 221

D
Digital Pocket Anemometer 176
Digital thermometer 175

G
Government strategic petroleum reserves (GSPRs) 159, 160

H
High-performance liquid chromatography (HPLC) 28

I
Intermittent Gas Lift (IGL) 221

K
Kegg Orthology (KO) 55

M
Mississippi Canyon Block 252 (MC252) 48

N
Non-linear algebraic equations (NLE) 232
Non-metric multidimensional scaling (nMDS) 59

O
Ordinary differential equations (ODE) 232

P
Pore pressure 1, 2, 3, 4, 5, 8, 9, 10, 11, 13, 14, 16, 19, 20

R

Renormalization group (RNG) 196, 203

S

Sediment cohesion 5
Small subunit (SSU) 53
Sol Carpenter H#7 (SC) 85
Spectral gamma ray (SGR) 113
Sugar Tree #1 (ST) 85
Sulfate- reducing bacteria (SRB) 24

T

Tensile stress 3, 11, 16

Total petroleum hydrocarbons (TPH) 58
Total petroleum hydrocarbon (TPH) 51
Transgressive surface of erosion (TSE) 113

U

United States Environment Protection Agency (USEPA) 194

X

X-ray diffraction (XRD) 27